Lessons Learned from the DESIGN, CONSTRUCTION *and* OPERATION *of* HYDROELECTRIC FACILITIES

Prepared by the Task Committee on Lessons Learned from the Design, Construction and Operation of Hydroelectric Facilities of the Hydropower Committee, Energy Division, of the American Society of Civil Engineers

PB POWER LTD
NEWCASTLE UPON TYNE

Published by the
American Society of Civil Engineers
345 East 47th Street
New York, New York 10017-2398

ABSTRACT

Lessons Learned from the Design, Construction and Operation of Hydroelectric Facilities is an addition to the series of publications by members of the Hydropower Committee of the American Society of Civil Engineers Energy Division. It provides information of civil facilities at hydroelectric projects based on experiences in designing, constructing, and operating of facilities and complements previous efforts by ASCE committees to disseminate information on lessons learned with dams and hydraulic structures. This encyclopedia of practical knowledge acquired by civil engineers through their design, construction, and operation of hydroelectric facilities presents the profession with many significant lessons in this area of civil engineering.

Library of Congress Cataloging-in-Publication Data

Lessons learned from the design, construction, and operation of hydroelectric facilities/prepared by the Task Committee on Lessons Learned from the Design, Construction, and Operation of Hydroelectric Facilities of the Hydropower Committee, Energy Division of the American Society of Civil Engineers.
    p.cm.
    Includes index.
    ISBN 0-7844-0000-8
    1. Hydroelectric power plants—Design and construction. 2. Hydroelectric power plants—Research—United States. I. American Society of Civil Engineers. Task Committee on Lessons Learned from the Design, Construction, and Operation of Hydroelectric Facilities.
TK1081.L47 1994   94-7387
621.31'2134—dc20   CIP

*Lessons Learned from the Design, Construction
and Operation of Hydroelectric Facilities*

FOREWORD

Lessons Learned from the Design, Construction and Operation of Hydroelectric Facilities is an addition to the series of publications by members of the Hydropower Committee of the American Society of Civil Engineers (ASCE) Energy Division. This document owes its conception to committee members who volunteered for a Task Force to determine, document, and disseminate professional lessons, both positive and negative, associated with the design, construction, and operation of hydroelectric facilities. The Task Force was formed at a meeting of the Hydropower Committee in Denver, Colorado, in July 1991. The specific features covered by this document include intakes/diversion, water conduits, power plants, and open channels. Lessons learned on dams, spillways, and other hydraulic structures are covered by other, existing ASCE publications.

The Hydropower Committee of ASCE's Energy Division was formed to **develop and disseminate information on all phases of hydroelectric power to the hydro engineering community.** The principal publication of the Hydropower Committee is *Civil Engineering Guidelines for Planning and Designing Hydroelectric Developments* (1989). This document provides supplemental information of civil facilities at hydroelectric projects based upon experiences in designing, constructing, and operating of facilities and complements previous efforts by ASCE committees to disseminate information on lessons learned with dams and hydraulic structures. The previous related volumes are *Lessons from Dam Incidents* and *Lessons from Dam Incidents USA - II* by the U.S. Committee on Large Dams (USCOLD) and published by ASCE and *Lessons Learned from the Design, Construction and Performance of Hydraulic Structures* by the Hydraulics Division of ASCE. These publications served as a guide for the collection of information by questionnaires.

The members of the Task Force thank the individuals and organizations that responded to the questionnaires and shared their experiences. The sharing of professional experiences on hydroelectric facilities through publications benefits the professional community and provides information that serves the public interest.

Lessons Learned from the Design, Construction
and Operation of Hydroelectric Facilities

ACKNOWLEDGEMENTS

ASCE Energy Division Executive Committee Contact Member

Arvids Zagars, P.E. (1993) Harza Engineering Company
Chicago, Illinois

ASCE Hydropower Development Committee Control Group Members

Dick Stutsman, P.E.   Ensigh & Buckley
152 Hilldale Dr.
San Enselmo, CA 94960

Ashok Rajpal, P.E.   Mead & Hunt, Inc.
6501 Watts Road
Madison, WI 53719

Antonio Ferreira, P.E.   6 Jeane Dr.
Holyoke, MA 01040

Ralph R. Rosa, Jr., P.E. U.S. Army Corps of Engineers
215 North 17th ST
Omaha. NE 68102

*Lessons Learned from the Design, Construction
and Operation of Hydroelectric Facilities*

ACKNOWLEDGEMENTS
(CONTINUED)

**Task Committee on Lessons Learned from the Design, Construction
and Operation of Hydroelectric Facilities**

Edward F. Carter, P.E. - Chairman
Harza Engineering Company
Sears Tower
233 South Wacker Drive
Chicago, IL 60606-6392

Bruce Ainsworth, P.E. - Vice Chairman
Black & Veatch Engineers
6 Venture, Suite 315
Irvine, CA 92718-3317

E. Richard Colle, P.E.
Geoconstruction Consultant
25 North Drive
Haddonfield, NJ 08033

Jerry S. Dodd, P.E.
Consulting Engineer and Geologist
5884 South Hannover Way
Englewood, CO 80111-3734

Paul J. Flanagan, P.E.
California Dept. of Water Resources
31770 West Highway 152
Santa Nella, CA 95322

Thomas G. Gebhard, Jr., Ph.D., P.E.
Gebhard●Sarma Group, Inc.
511 West 7th Street
Austin, TX 78701

Dan Pegan, P.E.
Pacific Gas & Electric Company
3400 Crow Canyon Road
San Ramon, CA 94538

Michael F. Rogers, P.E.
Harza Engineering Company
Sears Tower
233 South Wacker Drive
Chicago, IL 60606-6392

David E. Zehner, P.E.
Federal Energy Regulatory Commission
825 North Capitol St., NE
Washington, DC 20426

*Lessons Learned from the Design, Construction
and Operation of Hydroelectric Facilities*

ACKNOWLEDGEMENTS
(CONTINUED)

Technical Review Committee

1. Daniel Mahoney   Federal Energy Regulatory Commission

2. Bob E. Oxendine   Tapoco, Inc.

3. Linda Hinseth   Wisconsin Power & Light Company

4. David E. Jones   Yadkin, Inc.

5. Richard T. Gallus   Pennsylvania Electric Company

6. John A. Wilson   U.S. Bureau of Reclamation

7. Joseph W. Kick   Wisconsin Electric Power Company

8. Carl Vansant   Hydro Review Magazine, HCI Publications

*Lessons Learned from the Design, Construction
and Operation of Hydroelectric Facilities*

ACKNOWLEDGEMENTS
(CONTINUED)

Organizations that Provided Continuous Support for Committee Member Participation in Preparation of this Document

Harza Engineering Company   (Support for all administrative activities of the chairman)

Stone & Webster Engineering   (Support for arranging committee meetings)

Black & Veatch Engineers

Pacific Gas & Electric Company

Jerry S. Dodd, Consultant

California Department of Water Resources

Gebhard Engineers, Inc.

Federal Energy Regulatory Commission

*Lessons Learned from the Design, Construction
and Operation of Hydroelectric Facilities*

ABBREVIATIONS

ASCE	American Society of Civil Engineers
cfs	Cubic Feet per Second
cfm	Cubic Feet per Minute
EAP	Emergency Action Plan
FERC	Federal Energy Regulatory Commission
fps	Feet Per Second
HDPE	High density polyethylene
HP	Horsepower
HVAC	Heating, Ventilation and Air Conditioning
NDT	Non-destructive Testing
O&M	Operation and Maintenance
PS	Pumped-Storage
psi	Pounds per Square Inch
PUD	Public Utility District

Lessons Learned from the Design, Construction
and Operation of Hydroelectric Facilities

CONTENTS

| Lessons Learned from the Design, Construction |
| and Operation of Hydroelectric Facilities |

CHAPTER 1 - Introduction

1.1 Objective

Practical knowledge acquired by civil engineers through the design, construction, and operation of hydroelectric facilities has little value if it remains in office files or in the minds of the engineers who designed or built the projects. Predicting physical behavior by inference is as important as numerical analysis in civil engineering. Therefore, experience is a great educator in engineering, but it is impossible for an individual engineer during a career to be associated personally with many different hydroelectric works located in a variety of physical settings. Also, most engineers do not work for very large engineering firms where practical knowledge and experience can be exchanged informally by personal contact. Therefore, encyclopedias of significant lessons learned in civil engineering are important to the profession.

1.2 Scope

This document is a continuation of the effort by the American Society of Civil Engineers (ASCE) and the United States Committee on Large Dams (USCOLD) to record lessons from hydroelectric projects. Previous publications are titled "Lessons from Dam Incidents", "Lessons from Dam Incidents - II", and "Lessons Learned from the Design, Construction, and Performance of Hydraulic Structures." To complement these publications, the features covered in this document are the intakes/diversion, water conduits, power plants, and open channels.

The information was collected by sending a questionnaire to individuals and firms who are associated with hydroelectric developments. The responses are formatted and sorted by feature and the development stage to which the lesson pertains. Because of the number of responses to this initial endeavor, no attempt was made to evaluate collectively the responses by the contributors - each event stands on its own. It is hoped that over time this compendium will be expanded and that a technical evaluation of a number of similar cases may provide an interesting and worthwhile contribution to engineering aspects of hydroelectric developments. Engineers and engineering firms are encouraged to establish "Lessons" files on an appropriate form so that important knowledge and experience can be shared in the future through updating of this and similar documents.

CHAPTER 2 - Summary of Findings

2.1 Questionnaire Format

The format of the questionnaire is shown in **Appendix A**. The questionnaires were designed to be filled out quickly with only essential information on the structure, owner, and circumstances surrounding the lesson learned. Readers are encouraged to consult with the contact person for detail information on the problem and solution.

2.2 Questionnaire Response Summary

A total of 72 responses were received. One was rejected for not applying directly to hydroelectric facilities. The remaining 71 responses were classified according to project feature and stage as shown on Table 1.

TABLE 1 - Summary of Responses

PROJECT FEATURE	PROJECT STAGE		
	Design	Construction	Operation
Intakes/Diversion	0	0	11
Water Conduits	1	1	6
Power Plants	7	2	29
Open Channels	3	0	11

A total of 56 projects are represented by the questionnaires returned. A list of projects documented in the questionnaires is attached in **Appendix B**.

The information provided on the questionnaire was compiled into a similar format for all questionnaires returned.

CHAPTER 3 - Key Word Index of Abstracts by Project Feature and Stage

3.1 INTAKES/DIVERSION

3.1.1 Design

No abstracts returned.

3.1.2 Construction

No abstracts returned.

3.1.3 Operation

Index	Project/Owner	Keyword(1)	Keyword(2)	Keyword(3)
IO-001 p.4-1	Goodyear Lake Hydro Hydro Development Group, Inc.	Debris	Trashrack	Logboom
IO-002 p.4-2	Garvin Falls Hydro Pub. Service of N.H.	Debris	Trashrack	Ice
IO-003 p.4-3	Great Falls Hydro City of Patterson	Debris	Trashrack	Ice
IO-004 p.4-4	Wissota Hydro Dam Northern States Power	Gates	Dewatering	Maintenance
IO-005 p.4-5	Democrate Dam/Kern River 1 Southern California Edison	Drain	Inoperable	Gate
IO-006 p.4-6	Balsam Meadow P.S. Southern California Edison	Fish	Entrainment	
IO-007 p.4-7	Lower Monumental L & D Corps of Engineers	Uplift	Drains	Cold Temp
IO-008 p.4-8	New Martinsville Project City of New Martinsville, OH	Debris	Disposal	Recycling

3.1.3 Operation (continued)

Index	Project/Owner	Keyword(1)	Keyword(2)	Keyword(3)
IO-009 p.4-9	Holtwood Hydro Station Pennsylvania Power & Light	Stoplogs	Leakage	Neoprene
IO-010 p.4-10	Toledo Bend Project Sabine River Authority of TX Sabine River Authority of LA	Debris	Logboom	Crane
IO-011 p.4-11	Kansas River Project Bowersock Mills & Power Co.	Trashrack	Frazil	Ice

3.2 WATER CONDUITS

3.2.1 Design

Index	Project/Owner	Keyword(1)	Keyword(2)	Keyword(3)
WD-001 p.4-12	Bath County P.S. Virginia Electric & Power	Leakage	Grouting	Instrumentation

3.2.2 Construction

Index	Project/Owner	Keyword(1)	Keyword(2)	Keyword(3)
WC-001 p.4-13	Falls River Hydroelectric Project Marysville Hydro Partners	Compaction	Hydrostatic	Pressure

3.2.3 Operation

Index	Project/Owner	Keyword(1)	Keyword(2)	Keyword(3)
WO-001 p.4-14	Lundy Lake Dam Southern California Edison	Corrosion	Leakage	Spincoating
WO-002 p.4-15	Chief Joseph Dam Corps of Engineers	Baker Coupling	Movement	Leakage

3.2.3 Operation - continued

Index	Project/Owner	Keyword(1)	Keyword(2)	Keyword(3)
WO-003 p.4-16	Birch Creek Project Birch Power Co. & Sorenson Eng.	Radiograph	Weld	Testing
WO-004 p.4-17	Bishop Plant 2 Southern California Edison	Wood	Bell & Spigot	Leakage
WO-005 p.4-18	White River Project Northern States Power Co.	Wood Stave	Dewatering Collapse	Hoops
WO-006 p.4-19	Big Creek 2 Southern California Edison	Surge Tank	Dewatering	Drainage

3.3 POWER PLANTS

3.3.1 Design

Index	Project/Owner	Keyword(1)	Keyword(2)	Keyword(3)
PD-001 p.4-20	Murray Hydroelectric Project City of Little Rock, AR	Sheetpile	Vibration	Connections
PD-002 p.4-21	Barkley Power Plant Corps of Engineers	Spillway	Chains	Lubrication
PD-003 p.4-22	Northfield Mountain P.S. Northeast Utilities	Spherical	Valve	Seals
PD-004 p.4-23	Crystal Dam US Bureau of Reclamation	Water-hammer	Vortex	Venting
PD-005 p.4-24	Dillon Dam Denver Water Department	Turbine	Speed	Governor
PD-006 p.4-25	Lewisville Dam City of Denton, TX	Carbon	Seals	Flexing
PD-007 p.4-26	St. Anthony Falls Lower Dam Northern States Power Co.	Sandstone	Foundation	Piping

3.3.2 Construction

Index	Project/Owner	Keyword(1)	Keyword(2)	Keyword(3)
PC-001 p.4-27	Pit 3 Fishwater Release Pacific Gas & Electric Co.	Fish	Welds	NDT
PC-002 p.4-28	Holtwood Hydro Station Pennsylvania Power & Light	Reactive	Aggregate	Movement

3.3.3 Operation

Index	Project/Owner	Keyword(1)	Keyword(2)	Keyword(3)
PO-001 p.4-29	Upriver Dam Hydroelectric Project City of Spokane, WA	Load Rejection	EAP	Redundancy
PO-002 p.4-30	Maxwell Kohler, Sunshine, Oradell & Betasso Plants City of Boulder	Gates	Diversion	Generators
PO-003 p.4-31	Lloyd Shoals Hydro Plant Georgia Power Co.	Flashboards	Corrosion	Leakage
PO-004 p.4-32	Cuero Plant Cuero Hydro Partnership	Foundation	Settlement	Piping
PO-005 p.4-33	Toledo Bend Project Sabine River Authority/TX Sabine River Authority/LA	Erosion	Draft Tube	Lining
PO-006 p.4-34	Cherokee Falls Hydro Broad River Electric Coop.	Flooding	Flashboards	Debris
PO-007 p.4-35	Ellis Hydroelectric Plant Arkansas Electric Coop.	Corrosion	Stainless Steel	Pitting
PO-008 p.4-36	Colebrook Hydro Project MDC	Computer	Virus	Software
PO-009 p.4-37	Cougar Dam Corps of Engineers	Wicket Gate	Galling	Spatter
PO-010 p.4-38	Curecanti Morrow Pt. Dam US Bureau of Reclamation	Pump	Mineral	Deposits

3.3.3 Operation (continued)

Index	Project/Owner	Keyword(1)	Keyword(2)	Keyword(3)
PO-011 p.4-39	Glen Canyon Dam US Bureau of Reclamation	HVAC	Access	Platforms
PO-012 p.4-40	Glen Canyon Dam US Bureau of Reclamation	Generator	Rotor	Inspection
PO-013 p.4-41	Glen Canyon Dam US Bureau of Reclamation	Valve	Access	Platforms
PO-014 p.4-42	Holtwood Dam Pennsylvania Power & Light	Random	Vibration	Vortex Shedding
PO-015 p.4-43	Holtwood Dam Pennsylvania Power & Light	Dewatering	Generators	Wheel Pit
PO-016 p.4-44	Hoover Dam US Bureau of Reclamation	Generators	Controls	Air Bubble
PO-017 p.4-45	Lakeport Hydro Project NH Water Res. Council/ Lakeport Hydro Corp.	Power Sales	Automation	Controls
PO-018 p.4-46	Libby Dam Corps of Engineers	Runout	Vibration	Generator
PO-019 p.4-47	Ocoee III Hydro Plant Tennessee Valley Authority	Thrust Bearing	Sole Plates	Foundation Bolts
PO-020 p.4-48	Ray Roberts Dam City of Denton, TX	Equalizer	Turbine	Orifice
PO-021 p.4-49	Roanoke Rapids/Gaston Hydro Station Virginia Electric & Power	Cavitation	Overlays	Stainless Steel
PO-022 p.4-50	Dos Amigos Pumping Plant CA Dept. of Water Res.	Wear Rings	Pumps	
PO-023 p.4-51	Dos Amigos Pumping Plant CA Dept. of Water Res.	Rotors	Motors	Laminations
PO-024 p.4-52	B.F. Sisk-San Luis Dam CA Dept. of Water Res.	Amortissuer	Windings	Motors

3.3.3 Operation (continued)

Index	Project/Owner	Keyword(1)	Keyword(2)	Keyword(3)
PO-025 p.4-53	B.F. Sisk-San Luis Dam CA Dept. of Water Res.	Keys	Motors	Rotors
PO-026 p.4-54	B.F. Sisk-San Luis Dam CA Dept. of Water Res.	Resistance Welder	Windings	Motor
PO-027 p.4-55	B.F. Sisk-San Luis Dam CA Dept. of Water Res.	Switchyard	Breakers	Earthquake
PO-028 p.4-56	B.F. Sisk-San Luis Dam CA Dept. of Water Res.	Wedges	Rotor	Motors
PO-029 p.4-57	Wanapum Dam PUD No. 2 Grant Co.	Bolt Tension	Turbines	Packing Boxes

3.4 OPEN CHANNELS

3.4.1 Design

Index	Project/Owner	Keyword(1)	Keyword(2)	Keyword(3)
OD-001 p.4-58	Toledo Bend Project Sabine River Authority/TX Sabine River Authority/LA	Movement	Surveillance	
OD-002 p.4-59	Hatfield Dam Northern States Power Co.	Overtopping	Armoring	Drainage
OD-003 p.4-60	Kootenay Canal BC Hydro	Debris	Liner	

3.4.2 Construction

No abstracts submitted.

3-6

3.4.3 Operation

Index	Project/Owner	Keyword(1)	Keyword(2)	Keyword(3)
OO-001 p.4-61	Oswegatchie Development Niagara Mohawk Power	Wind	Connectors	Corrosion
OO-002 p.4-62	Bishop 2 Hydro Project Southern California Edison	Dredging	Suction	Sediment
OO-003 p.4-63	Sherman Island Dev. Niagara Mohawk Power	Freeze-Thaw	Soil	Insulation
OO-004 p.4-64	Toledo Bend Project Sabine River Authority/TX Sabine River Authority/LA	Riprap	Scour	
OO-005 p.4-65	Self Cleaning Weir Pacific Gas & Electric	Sediment	Weirs	Instrumentation
OO-006 p.4-66	Kaweah 3 Project Southern California Edison	Canal	Leakage	Shotcrete
OO-007 p.4-67	Lee Vining Substation Southern California Edison	Fisheries	Creeks	Gunite
OO-008 p.4-68	B.F. Sisk-San Luis Dam CA Dept. of Water Res.	Leaks	Divers	Grout Holes
OO-009 p.4-69	B.F. Sisk-San Luis Dam CA Dept. of Water Res.	Cofferdam	Canal	Turnouts
OO-010 p.4-70	Thomson Project Minnesota Power Co.	Rapid Drawdown	Dewatering	Slope Failure
OO-011 p.4-71	White River Project Puget Sound Power & Light	Timber Lined	Piping	

CHAPTER 4 - Abstracts by Project Feature and Stage

4.1 INTAKES

Project:	Goodyear Lake Hydro	Index No:	IO-001
Owner:	Hydro Development Group, Inc.	Feature:	Intake
		Stage:	Operation
River/Stream:	Goodyear Lake, NY	Keyword(1):	Debris
Capacity:	4.585 MW	Keyword(2):	Trashrack
Nearest City/State:		Keyword(3):	Logboom
Contact Name:	Mr. Mark E. Quallen		
Contact Title:	President		
Address:	P.O. Box 58		
City/State/Zip:	Dexter, NY 13634		
Telephone:	(315) 639-6700		

LESSON LEARNED

Problem: Operation of power plant was not economical when debris clogged trashrack.

Cause & Effect: Debris build up on intake trashrack was causing restricted flow.

Investigations: N/A

Analysis/Action: Installed trashrack rake and log boom in front of power canal.

Lessons Learned: Before completing design of any intake structure, a study must be made to estimate the volume of debris to be handled, debris impacts on generation, and determination of procedures to prevent debris build up.

Publications: None.

Project:	Garvin Falls Hydro	Index No:	IO-002
Owner:	Public Service Company of New Hampshire	Feature: Stage:	Intake Operation
River/Stream: Capacity: Nearest City/State:	 17.222 MW New Hampshire	Keyword(1): Keyword(2): Keyword(3):	Debris Trashrack Ice
Contact Name: Contact Title:	J. E. Lyons N/A		
Address: City/State/Zip: Telephone:	1000 Elm Street Manchester, NH 03105 (603) 669-4000		

LESSON LEARNED

Problem: Excessive year-round debris build up on trashrack and unmanageable ice jams in the intake canal in winter.

Cause & Effect: The installation of two new generating units (additional capacity) significantly increased intake canal velocities, causing build up of debris and ice jams - which at times caused the complete shut down of the station so that the debris and ice could be removed.

Investigations: A two year study was performed.

Analysis/Action: Remedial action consisted of installing of automatic trashrack and manually controlled, hydraulic articulated crane to remove debris and ice.

Lessons Learned: Complete study of effects on operations of facility is needed before installation of new capacity/units.

Publications: DOE/ID/12122,1 NTIS Publications
DOE "Small-scale hydropower Program- Feasibility Assessment and Technical Development " Final Report DOE/ID-10322

Project:	Great Falls Hydro	Index No:	IO-003
Owner:	City of Patterson	Feature:	Intake
		Stage:	Operation
River/Stream:	Passaic River	Keyword(1):	Debris
Capacity:	119.292 MW	Keyword(2):	Trashrack
Nearest City/State:	Patterson, NJ	Keyword(3):	Ice
Contact Name:	J. Topalian		
Contact Title:			
Address:	52 Church Street		
City/State/Zip:	Patterson, NJ 07505		
Telephone:	(210)881-3313		

LESSON LEARNED

Problem: Debris and frazzle ice built up causing the intake trashrack to collapse.

Cause & Effect: The debris and ice completely blocked the trashrack. The water gages used to determine the differential pressure were not functioning; a 100% blockage of the rack caused the rack to collapse. This required the replacement of the rack and resulted in lost generation.

Investigations: Investigations were found to be too expensive.

Analysis/Action: A new trashrack designed for 100% blockage was installed together with a power operated trash rake.

Lessons Learned: Design review of small projects before construction is warranted to prevent similar failures. Trashracks should be designed for 100% blockage. Debris removal equipment and water differential alarms should be secondary line of defense to preventing failure of trash rack.

Publications: .DOE/ID/12127-1,2 NTS Publications, DOE "Small Scale Hydropower Program-Feasibility Assessment and Technology Development" Final Report

Project:	Wissota Hydro Dam	Index No:	IO-004
Owner:	North States Power Company	Feature:	Intake
		Stage:	Operation
River/Stream:	Chippewa River	Keyword(1):	Gates
Capacity:	36 MW	Keyword(2):	Dewatering
Nearest City/State:	Chippewa Falls, Wisconsin	Keyword(3):	Maintenance
Contact Name:	Richard Rudolph		
Contact Title:			
Address:	PO Box 8		
City/State/Zip:	Eau Claire, WI 54702		
Telephone:			

LESSON LEARNED

Problem: The existing seventy-year old stoplog embedded metalwork, trashracks and rack supports needed to be replaced. The only means of repairing or replacing these items was to dewater the intake by lowering the reservoir.

Cause & Effect: The problem resulted from long term corrosion of the submerged steel. This corrosion was accelerated by the loss of the protective steel coatings . It was found that some of the critical members were reduced in section by as much as 70% due to the corrosion. Because these structures were located upstream of the nearest point for dewatering (the bulkhead gate) no prior means for dewatering the intake had been established.

Investigations: An investigation was performed by divers who provided a condition survey and verified the structural dimensions. A structural evaluation was also completed to determine the load bearing capacity

Analysis/Action: To perform the replacement work conventional bulkheads using barge and crane, underwater replacement and floating bulkheads were considered. The floating bulkhead scheme was selected based on lowest cost, ease of operation, and capability for use at other facilities.

Lessons Learned: A vertical or near vertical sloping face and sill should be provided on intake structures upstream of submerged steel elements susceptible to corrosion. This surface provides the capability of installing a temporary bulkhead so that repairs can be performed. To extend the life of submerged structures, it is recommended that non-corrosion materials be considered.

Publications: 1) "Floating Bulkhead Proves Flexible, Reusable", K. Sirotiar, Hydro Review, June 1987.

2) "An Innovative Concept for Dewatering Hydro Plants", Frederick Lux III and James R. Bakken, Hydro Review, Volume XI Number 7, December 1992.

3) "Considerations for Dewatering Hydro Units and Gates", R. M. Rudolph and J. Kries, Waterpower, Aug 10-13, 1993.

Project:	Democrate Dam/Kern River 1	Index No:	IO-005
	Intake	Feature:	Intake
Owner:	Southern California Edison	Stage:	Operation
River/Stream:	Kern River	Keyword(1):	Drain
Capacity:	24.8 MW	Keyword(2):	Inoperable
Nearest City/State:		Keyword(3):	Gate

Contact Name:	G. B. Redd
Contact Title:	
Address:	PO Box 800
City/State/Zip:	Rosemead, CA 91770
Telephone:	(818) 302-8950

LESSON LEARNED

Problem: The drain tunnel for the intake was originally equipped with a upper and lower shutoff valve. The upper gate had deteriorated so that is was inoperable and the lower gate required a large force to open and close. There was a concern that the lower gate would fail and render the intake unusable.

Cause & Effect: Fear of the lower gate failure resulted in discontinued use of the drain tunnel and siltation of the intake. The siltation made it difficult to regulate the intake pond which posed a risk during an oil spill.

Investigation: Alternatives to re-establish use of the drain tunnel and reliability of the system were studied and evaluated.

Analysis/Action: Replacement of the existing gates would drain the system and was not the preferred alternative. The best alternative was to construct a new concrete gate structure downstream of the lower gate and replace old lower gate hydraulic operators with new mechanical operators. In this alternative construction was possible without affecting operations because the lower gate served as guard gate . This alternative still uses the lower gate as a back-up and flushing can now be routinely performed.

Lessons Learned: It is sometimes better to abandon old equipment in place rather than attempting to repair or modify it. Old hydraulic operators can be replaced by mechanical operators.

Publications: None.

Project:	Balsam Meadow Pumped Storage	Index No:	IO-006
Owner:	Southern California Edison	Feature:	Intake
		Stage:	Operation
River/Stream:	Shaver Lake	Keyword(1):	Fish
Capacity:	207 MW	Keyword(2):	Entrainment
Nearest City/State:	Shaver Lake City	Keyword(3):	Cold Temp

Contact Name:	S. F. McKenery
Contact Title:	Project Manager
Address:	PO Box 800, Bldg GO-3
City/State/Zip:	Rosemead, CA 91770
Telephone:	(818) 302-8572

LESSON LEARNED

Problem: The Federal Energy Regulatory Commission (FERC) requires a detailed study of operational impact on the Shaver Lake fish habitat.

Cause & Effect:

Investigation: The project went into conventional operations in 1987 and was modified to a pumped storage facility in 1990. The required fish entrainment study was performed from July 1991 to July 1992. After evaluating various methods it was decided to use hydro acoustic probes across the intake structure 100% of the time with select netting across the discharge three days each month to validate the acoustic measurements.

Analysis/Action: The equipment was installed and data collected for one year. The study found that there was minor fish entrainment from Shaver Lake during pumped storage operations.

Lessons Learned: Hydro acoustic monitoring is effective, a report has been submitted to FERC for its concurrence.

Publications: The Acoustic Monitoring final report has been approved by the local agencies.

Project:	Lower Monumental Lock and Dam	Index No:	IO-007
Owner:	U. S. Army Corps of Engineers	Feature:	Intake
		Stage:	Operation
River/Stream:	Snake River	Keyword(1):	Uplift
Capacity:	810 MW	Keyword(2):	Drains
Nearest City/State:	Pasco, WA	Keyword(3):	Cold Temps
Contact Name:	William Harrison		
Contact Title:	Staff Geologist		
Address:	Walla Walla District		
City/State/Zip:	Walla Walla, WA		
Telephone:	(509) 522-6767		

LESSON LEARNED

Problem: Closed cell piezometers located under a powerhouse intake slab indicated pressures exceeding the design uplift for the slab.

Cause & Effect: There is no known cause for the uplift. An unknown water source under the intake slab was causing uplift pressure above design uplift values.

Investigation: A study was undertaken to determine the source of the water causing increased uplift pressures. Also, corrective measures were taken to relieve uplift pressures. The work included cleaning existing foundation and horizontal drains by over reaming, installing a new closed cell piezometer to verify existing pressure readings, pressure tests by double packer on all drains and new piezometer hole, crack mapping in the drainage and grouting gallery, and temperature/pressure monitoring after completion.

Analysis/Action: After cleaning a single horizontal drain adjacent to the piezometer locations the uplift pressure began to drop. Water temperature measurements for forebay, drains, and piezometers indicated that the water source may be from the forebay rather than the foundation. This intern indicated that there may be a crack in the intake slab.

Lessons Learned: Temperature monitoring may be a good indicator of water source. Cleaning foundation drains (in this case) had no effect on the uplift pressures. Cleaning and/or installing horizontal drains may relieve pressures under intake structures in the case of cracks in the intake slab. The source of the cracks may require additional studies concerning structural stability.

Publications: None.

Project:	New Martinsville Project	Index No:	IO-008
Owner:	City of New Martinsville	Feature:	Intake
		Stage:	Operation
River/Stream:	Ohio River	Keyword(1):	Debris
Capacity:	34 MW	Keyword(2):	Disposal
Nearest City/State:	New Martinsville	Keyword(3):	Recycling
Contact Name:	David Pritchard		
Contact Title:	Project Engineer		
Address:	5085 Reed Road		
City/State/Zip:	Columbus, OH		
Telephone:	(614) 459-2050		

LESSON LEARNED

Problem: Intake debris (mostly wood) removal and disposal is a major expense.

Cause & Effect: The Ohio River carries a large quantity of floating and submerged debris which clogs the hydro facility intake and causes a reduction in generating capacity.

Investigation: None.

Analysis/Action: Instead of using conventional methods to dispose of the debris the Owner transports the wood to a handling area where it is cut and chipped for fuel and mulch. Local citizens then take the fuel and mulch away at no cost to the owner.

Lessons Learned: Wood debris disposal expense can be reduced by using wood for fuel and mulch. Redesign the intake to limit the accumulation of debris on the trashrack might also be tried. It would be of great interest if an intake that would limit trash accumulation on the Ohio River, at a reasonable cost, was designed.

Publications: None.

Project:	Holtwood Hydroelectric Station	Index No:	IO-009
Owner:	Pennsylvania Power & Light	Feature:	Intake
		Stage:	Operation
River/Stream:	Susquehanna	Keyword(1):	Stoplogs
Capacity:	102 MW	Keyword(2):	Leakage
Nearest City/State:	Lancaster, PA	Keyword(3):	Neoprene
Contact Name:	G. David Hopfer		
Contact Title:	Project Engineer		
Address:	2 N. 9th street, N5		
City/State/Zip:	Allentown PA 18101		
Telephone:	(215) 774-6816		

LESSON LEARNED

Problem: Steel headgates and draft tube stoplog timbers leaked excessively causing difficulties in dewatering the powerhouse.

Cause & Effect: The timbers quickly split forming passageways for water. In addition, the seal between the timbers and the steel frames is poor which allows for leakage. To limit the leakage the timbers frequently require replacement.

Investigation: Neoprene will last longer and provide a better seal than timbers.

Analysis/Action: A plate steel and neoprene arrangement was designed to replace the timbers. On-site construction forces installed the new design. The design consisted of two inch-thick 50 Durometer neoprene installed on a recessed steel plate.

Lessons Learned: The new arrangement is working well and has proven to be durable. The timber design was 83 years old making is necessary to persuade reluctant personnel to try new material. The success of neoprene as a water seal, and the high maintenance cost associated with timber, has prompted a program to eliminate timber as a water seal material.

Publications: None.

Project:	Toledo Bend Project Joint Operation	Index No:	IO-010
Owner:	Sabine River Authority of Texas and	Feature:	Intake
	Sabine River Authority of Louisiana	Stage:	Operation
River/Stream:	Sabine River	Keyword(1):	Debris
Capacity:	81 MW	Keyword(2):	Logboom
Nearest City/State:	Burkeville, Texas	Keyword(3):	Crane
	Anacoco, Louisiana		
Contact Name:	Bartom Rumsey		
Contact Title:	Project Engineer		
Address:	Rt. 1, Box 780		
City/State/Zip:	Many, LA 71449-9730		
Telephone:	(318) 256-4112		

LESSON LEARNED

Problem: The intake log boom was inadequate to contain the size and type of debris found in the Sabine River.

Cause & Effect: The debris, coupled with wind and wave conditions, continually broke the log boom cable allowing the debris to accumulate on the intake trashrack. The intake trash rake was inadequate for the size of debris that passed the log boom.

Investigation: None.

Analysis/Action: The boom was abandoned. The debris is removed from the intake by crane and clam shell. Over the years the volume of debris requiring removal by crane has reduced. Debris removal is now done about every 2 to 3 years.

Lessons Learned: Timber removal prior to inundation of the reservoir is the best way to address debris problems latter.

Publications: None.

Project:	Kansas River Project	Index No:	IO-011
Owner:	The Bowersock Mills & Power Co.	Feature:	Intake
		Stage:	Operation
River/Stream:	Kansas River	Keyword(1):	Trashrack
Capacity:	2.1 MW	Keyword(2):	Frazil
Nearest City/State:	Lawrence, KS	Keyword(3):	Ice
Contact Name:	Stephen H. Hill		
Contact Title:	President		
Address:	PO Box 66		
City/State/Zip:	Lawrence, KS 66044		
Telephone:	(913) 843-1356		

LESSON LEARNED

Problem: Ice would form on trashracks from water level to bottom of racks shutting off all water flow to turbines resulting in plant shutdown.

Cause & Effect: Operating personnel would permit headwater levels to fall exposing steel trash racks to cold air, and also thereby increasing the velocity of water flow in the forebay.

Investigations: Empirical studies plus several articles in Hydro Review led to the partial solution outlined below.

Analysis/Action: 1. Design trashracks for minimum exposure to air and keep water levels high. 2. Slow the rate of water flow through trash racks by design or by reducing gate settings on turbines. 3. Reduce plant operation until mill pond freezes over.

Lessons Learned: Frazil ice will not form if water flow is slow and especially if the water comes to plant from an ice covered mill pond, presumably because the water is warmer. Minimizing trashrack exposure is also key.

Publications: One or two articles have appeared in Hydro Review over last two or three years which touched on this problem.

Other Comments: Operating personnel have learned to curtail production, if necessary, for several days to allow ice formation on the mill pond. At a project's latitude ice may come and go several times each winter, causing difficult operating conditions and complete shutdowns.

4.2   WATER CONDUITS

Project:	Bath County Pumped Storage Station	Index No:	WD-001
Owner:	Virginia Electric and Power Company	Feature: Stage:	Water Conduit Design

River/Stream:	Back Creek	Keyword(1):	Leakage
Capacity:	2100 MW	Keyword(2):	Grouting
Nearest City/State:	Warm Springs, VA	Keyword(3):	Instrumentation

Contact Name:	Michael Wood
Contact Title:	Supervisor, Geotechnical & Environmental

Address:	HCR-01, Box 280
City/State/Zip:	Warm Springs, VA 24484
Telephone:	(703) 279-3204

LESSON LEARNED

Problem: Initial filling of power tunnel 1 resulted in failure of the adjacent empty penstock in power tunnel 2. Significant tunnel leakage was observed at the time of filling.

Cause & Effect: Higher than anticipated permeability of the rock mass. Hydrostatic head within the tunnel was greater than in situ stress.

Investigations: In-situ stress investigation was conducted.

Analysis/Action: High pressure (600 psi) grouting program combined with drainage and instrumentation program were implemented. Long-term geotechnical monitoring and drain hole maintenance program also were put in place.

Lessons Learned: More rigorous testing, analysis and evaluation of results.

Publications: Oechsel, R.G., D.T. Wafle, and K.L. Wong, "Bath County Pumped Storage Project Tunnel System, Evaluation of Remedial Measures," presented at ISRN Symposium, Madrid, Spain, September 12-16, 1988.

Other Comments: None.

Project:	Falls River Project	Index No:	WC-001
Owner:	Marysville Hydro Partners	Feature:	Water Conduit
		Stage:	Construction
River/Stream:	Fall River	Keyword(1):	Compaction
Capacity:	9.1 MW	Keyword(2):	Hydrostatic
Nearest City/State:	Ashton, Idaho	Keyword(3):	Pressure

Contact Name:	Blaine Graff
Contact Title:	Manager of Engineering
Address:	1199 Shoreline Lane, Suite 310
City/State/Zip:	Boise, ID 83702
Telephone:	(208) 336-4254

LESSON LEARNED

Problem: The Falls River penstock failed due to backfill saturation and piping by water from the Marysville irrigation canal.

Cause & Effect: The Marysville irrigation canal crossed over the partially constructed penstock, near the head end of the penstock. The canal was in operation for almost a month prior to the failure. Operation of the canal was to terminate upon completion of the Falls River project. Insufficient compaction of the penstock backfill allowed the water from the canal to pipe along the haunches of the penstock until it daylighted at the face of the trench backfill. The subsequent loss of backfill material allowed the penstock to buckle under the weight of the overburden. The buckling proceeded uphill to the canal, whereupon the canal bank failed because of the buckled penstock directly beneath. The failure washed up to 17,000 cubic yards of soil into the Fall River.

Investigation: A field investigation determined that operation of the canal, lack of isolation of the canal water from the penstock backfill, and inadequate compaction of the penstock backfill contributed to the failure.

Analysis/Action: An analysis indicated that the penstock buckled under external hydrostatic/soil pressure from the canal and its backfill.

Lessons Learned: Provide adequate backfill compaction for penstock, minimize external pressures, and leave bracing in place until facility goes into operation.

Publications: None.

Project:	Lundy Lake Dam	Index No:	WO-001
Owner:	Southern California Edison Co.	Feature:	Water Conduit
		Stage:	Operation
River/Stream:	Mill Creek	Keyword(1):	Corrosion
Capacity:	3 MW	Keyword(2):	Leakage
Nearest City/State:		Keyword(3):	Spincoating
Contact Name:	G. B. Redd		
Contact Title:			
Address:	PO Box 800		
City/State/Zip:	Rosemead, CA 91770		
Telephone:	(818) 302-8950		

LESSON LEARNED

Problem: A 5-foot-diameter discharge pipe which supplied water for agriculture needs was located in the face of a rockfill dam. The 200 foot long pipe was of riveted construction and badly corroded. The California State Division of Safety of Dams (DSOD) was concerned that the pipe would collapse and cause the dam failure.

Cause & Effect: The level of corrosion and general condition of the pipe was attributed to its age and leaking joints. The pipe varied in size and the joints were stepped making them difficult to seal.

Investigation: The pipe wall thickness and a structural analysis found the pipe to be acceptable for the existing loading.

Analysis/Action: It was determined that replacement of the pipe was not possible due to the dam construction type (rockfill). Three repair schemes were evaluated-steel sleeve, epoxy sleeve and cement spincoating; the best alternative was the cement spincoating. Before the pipe was coated the joints were hand packed.

Lessons Learned: Cement coating is applicable to all diameter piping and is acceptable to DSOD.

Publications: None.

Project:	Chief Joseph Dam	Index No:	WO-002
Owner:	U.S. Army Corps of Engineers	Feature:	Water Conduit
		Stage:	Operation
River/Stream:	Columbia	Keyword(1):	Baker Coupling
Capacity:	2,069 MW	Keyword(2):	Movement
Nearest City/State:	Bridgeport, WA	Keyword(3):	Leakage

Contact Name:	David Raisanen
Contact Title:	Structural Engineer, Hydroelectric Design Center
Address:	P.O. Box 2870
City/State/Zip:	Portland, OR 97208
Telephone:	(503) 326-4916

LESSON LEARNED

Problem: The penstock Baker coupling leaked excessively at units 17 through 27.

Cause & Effect: Temperature fluctuations caused penstock movement which resulted in the couplings leaking. On one occasion a 6 foot length of the coupling gasket blew out.

Investigation: The penstock is 25 feet in diameter and inclined at 44 degrees from horizontal with approximately 105 feet of head. On-site investigations were performed by engineering and operations personnel. A stress analysis was performed by the design staff.

Analysis/Action: To remedy the problem the following steps were taken: the coupling bolts were loosened then the couplings, follower rings and gaskets were repositioned, and the bolts were retightened according to a specified procedure.

Lessons Learned: Several lessons were learned; 1) The penstock shell should have been locally thicker at the coupling; 2) Temperature induced movement should be limited to 3/8" at this type of coupling 3) Installation must be closely monitored to ensure specified tightening procedures are met; and 4) This type of coupling should be designed to take differential axial movement of the penstock ends at both gaskets, rather than letting one end take all the movement.

Publications: None.

Project:	Birch Creek Project	Index No:	WO-003
Owner:	Birch Power Company & Sorenson Engineering	Feature:	Water Conduit
		Stage:	Operation
River/Stream:	Birch Creek	Keyword(1):	Radiograph
Capacity:	2.7 MW	Keyword(2):	Weld
Nearest City/State:	Idaho Falls, Idaho	Keyword(3):	Testing

Contact Name:	Ted Sorenson
Contact Title:	
Address:	550 Linden Drive
City/State/Zip:	Idaho Falls, ID 83401
Telephone:	(208) 522-8069

LESSON LEARNED

Problem: The Catastrophic failure of manufacturer's coil spliced butt weld for 51-inch-diameter spiral welded pipe resulted in the collapse of approximately 7,000 feet of the penstock, although factory welds had been pressure tested prior to shipping.

Cause & Effect: It was determined that insufficient weld strength of coil spliced butt weld led to the failure under load.

Investigations: Pipe stress modeling for transient flow conditions.

Analysis/Action: Same as above.

Lessons Learned: Pipe cannot be designed for a catastrophic event. Radiographic testing of manufacturer's coil spliced welds is important.

Publications: None.

Other Comments: None.

Project:	Bishop Plant 2	Index No:	WO-004
Owner:	Southern California Edison Co.	Feature:	Water Conduit
		Stage:	Operation
River/Stream:	Bishop Creek	Keyword(1):	Wood
Capacity:		Keyword(2):	Bell & Spigot
Nearest City/State:	Bishop, CA	Keyword(3):	Leakage
Contact Name:	S. F. McKenery		
Contact Title:	Project Manager		
Address:	P.O. Box 800, Bldg. GO-3		
City/State/Zip:	Rosemead, CA 91770		
Telephone:	(818) 302-8572		

LESSON LEARNED

Problem: This 10,000 foot long, 60" diameter conduit was constructed out of redwood staves and steel bands. It was leaking badly.

Cause & Effect: The redwood conduit was last replaced in the 1950's and was located at about 8,000 foot elevation. The harsh winters and hot summers had caused a lot of damage and leaks were prevalent everywhere. The conduit is located above a state highway; leaking water in winter caused severe ice patches on the roadway.

Investigations: Replacement in kind was not practical since wood would again deteriorate. Steel pipe was a more practical replacement.

Analysis/Action: The redwood line was replaced with a 1/4" thick spiral welded pipe.

Lessons Learned: The pipe uses bell and spigot welded joints with a bolted dresser coupling every fifth joint for expansion/contraction flexibility. Since this coupling is not physically attached to either piece of pipe, some "creep" has occurred over time. The next pipe replacement will utilize a similar clamp-on coupling which is physically restrained on one end, allowing some flexibility but limiting creep from both ends.

Publications: None.

Other Comments: None.

Project:	White River Project	Index No:	WO-005
Owner:	Northern States Power Company	Feature:	Water Conduit
		Stage:	Operation
River/Stream:	White River	Keyword(1):	Wood Stave
Capacity:	1.0 MW	Keyword(2):	Dewatering
Nearest City/State:	Ashland, WI		Collapse
		Keyword(3):	Hoops
Contact Name:	Richard Rudolph		
Contact Title:	Hydro Administrator		
Address:	100 N. Barstow Street, P. O. Box 8		
City/State/Zip:	Eau Claire, WI 54702-0008		
Telephone:	(715) 839-2486		

LESSON LEARNED

Problem: A portion of the wood stave pipeline collapsed.

Cause & Effect: The tainter gates were used to draw down the reservoir for seal repairs. As a result, the wood stave pipeline collapsed when the pressure dropped, apparently due to stress concentrations aggravated by earlier loss of some internal staves.

Investigations: A consulting engineer performed an inspection and reviewed plant maintenance records after the collapse.

Analysis/Action: The pipeline was replaced using reinforced concrete pipe.

Lessons Learned: Exercise caution and carefully monitor when dewatering pressurized conduits due to stress reversals and concentrations. Investigate hoops prior to dewatering to verify stress and conditions.

Publications: None.

Other Comments: Relaxation of external steel hoops of the pipeline may have allowed the timber pipe to deform when the internal pressure was relieved. Instability occurs when the wood stave section deviates from its circular shape. Tightening the hoops prior to dewatering may have prevented excessive deformation.

Project:	Big Creek 2	Index No:	WO-006
Owner:	Southern California Edison Co.	Feature:	Water Conduit
		Stage:	Operation
River/Stream:	San Joaquin River	Keyword(1):	Surge Tank
Capacity:		Keyword(2):	Dewatering
Nearest City/State:	Shaver Lake, CA	Keyword(3):	Drainage
Contact Name:	S. F. McKenery		
Contact Title:	Project Manager		
Address:	P.O. Box 800, Bldg. GO-3		
City/State/Zip:	Rosemead, CA 91770		
Telephone:	(818) 302-8572		

LESSON LEARNED

Problem: The 9-feet-diameter penstock pipe cracked immediately downstream from the surge chamber of Big Creek Powerhouse 2.

Cause & Effect: The pipe became buoyant inside the tunnel following dewatering for a tunnel inspection.

Investigations: Inspection of the pipe inside and out was made immediately after the cracking was noticed to determine the cause of the break and to recommend repairs.

Analysis/Action: In addition to repairing the crack, a new drainage pipe was added to the lower end of the tunnel to prevent the tunnel from filling with water around the pipe.

Lessons Learned: Tunnels which have pipes in them should be properly drained and drains maintained on a routine basis.

Publications: None.

Other Comments: None.

4.3 POWER PLANTS

Project:	Murray Hydroelectric Project	**Index No:**	PD-001
Owner:	City of Little Rock, AR	**Feature:**	Power Plants
		Stage:	Design
River/Stream:	Arkansas River	**Keyword(1):**	Sheetpile
Capacity:	36.8 MW	**Keyword(2):**	Vibration
Nearest City/State:	North Little Rock, AR	**Keyword(3):**	Connections
Contact Name:	Mrs. Catharine Wilkins		
Contact Title:	General Manager		
Address:	PO Box 159		
City/State/Zip:	North Little Rock, AR 72115		
Telephone:	(501) 372-0100		

LESSON LEARNED

Problem: The powerhouse intake walls are constructed of sheetpiles with wall sections connected to the sheet piles. The connection between the wall and sheetpiles becomes loose periodically.

Cause & Effect: At this time it is thought that either load rejection vibration or normal operating equipment vibrations are causing the connections to loosen.

Investigation: The owner has retained a consulting engineer to perform an investigation. At this time the investigation results are not available.

Analysis/Action: During the investigation, the owner has decided to improve the connections and increase the sheetpile stiffness.

Lessons Learned: Design analysis for all operating load cases should be performed.

Publications: None.

Project:	Barkley Power Plant	Index No:	PD-002
Owner:	U.S. Army Corps of Engineers	Feature:	Power Plants
		Stage:	Design
River/Stream:	Cumberland River	Keyword(1):	Spillway
Capacity:	130 MW	Keyword(2):	Chains
Nearest City/State:	Gilbertsville, KY	Keyword(3):	Lubrication
Contact Name:	Jack Rowland		
Contact Title:	Mechanical Engineer, Hydropower Branch		
Address:	P.O. Box 1070		
City/State/Zip:	Nashville, TN 37202-1070		
Telephone:	(615) 736-5868		

LESSON LEARNED

Problem: Spillway gate lifting chain link pins were not provided with grease passages and were not sized so that grease passages could added.

Cause & Effect: The problem was not addressed during design has resulted in severe galling of bearing surfaces and subsequent inoperability.

Investigation: It was determined that the chain links became inflexible and would not follow the guides or conform to drive sprocket curvature, subsequently the guards and guides were damaged.

Analysis/Action: Chains were disassembled, pins ground and links bored. Spare sets of chains were purchased. Grease was worked partially into the bearing surface.

Lessons Learned: The problem persists and remedial actions continue. Chain drives should be sized sufficiently for lubrication or an alternative lifting means designed.

Publications: None.

Project:	Northfield Mountain Pump Storage Project	Index No:	PD-003
Owner:	Northeast Utilities	Feature:	Power Plants
		Stage:	Design
River/Stream:	Conn.	Keyword(1):	Spherical
Capacity:	1000 MW	Keyword(2):	Valve
Nearest City/State:	Northfield, MA	Keyword(3):	Seals
Contact Name:	Charles McKee		
Contact Title:			
Address:			
City/State/Zip:			
Telephone:	(509) 663-8121		

LESSON LEARNED

Problem: Flooded Powerhouse.

Cause & Effect: During initial filling of the upper reservoir the seals on spherical valve opened unintentionally, because of an inadvertent closure of a drain valve.

Investigation: Investigation by both Stone & Webster and Northwest Utilities have been performed

Analysis/Action: It was determined that the connection between the upstream seal and downstream seal should not have existed.

Lessons Learned: Do not parallel independent hydraulic systems, if it is done they are no longer independent.

Publications: None.

Project:	Colorado River Storage -	Index No:	PD-004
	Crystal Dam	Feature:	Power Plants
Owner:	US Bureau of Reclamation	Stage:	Design
River/Stream:	Gunnison River	Keyword(1):	Waterhammer
Capacity:	28 MW	Keyword(2):	Vortex
Nearest City/State:	Montrose, CO	Keyword(3):	Venting

Contact Name:	Jack Sage
Contact Title:	Maintenance Superintendent, CCI Field Division
Address:	1820 South Rio Grande Avenue
City/State/Zip:	Montrose, CO 81401
Telephone:	(303) 240-6306

LESSON LEARNED

Problem: Severe turbine draft tube water hammer.

Cause & Effect: Water vortexes in draft tube, limits operating loads of generator.

Investigations: Design of draft tube allowed two vortex's to develop in tube A at certain water flows, 600-1300 cfs. The periodic collapse of a vortex caused water hammer.

Analysis/Action: Venting system designed to stop vortex's from forming did not stop problem. Injecting compressed air (60 GFM) under turbine head cover completely stopped vortex's from forming.

Lessons Learned: None.

Publications: None.

Project:	Dillon Dam	Index No:	PD-005
Owner:	Denver Water Department	Feature:	Power Plants
		Stage:	Design
River/Stream:	Dillon Reservoir/Blue River	Keyword(1):	Turbine
Capacity:	1.75 MW	Keyword(2):	Speed
Nearest City/State:	Dillon, CO	Keyword(3):	Governor
Contact Name:	David C. Stone		
Contact Title:	Chief Mechanical Engineer		
Address:	1600 W. 12th Avenue		
City/State/Zip:	Denver, CO 80254		
Telephone:	(303) 628-6655		

LESSON LEARNED

Problem: During initial start-up, the turbine/generator was unstable; the governor could not adequately control the unit speed for synchronization.

Cause & Effect: The ratio of machine inertia to water inertia was below the minimum required by the governor, making it difficult for the governor to control speed. This was a result of miscommunication between turbine and generator manufacturers.

Investigations:

Analysis/Action: Governor manufacturer fine-tuned governor to its limits, leaving no room for additional adjustment or misadjustment due to wear. This created concern for future operation as unit experiences wear.

Lessons Learned: Owner should specify minimum machine/water inertia ratio to match specified governor requirements to insure unit stability.

Publications: None.

Other Comments: None.

Project:	Lewisville Dam	Index No:	PD-006
Owner:	City of Denton, TX	Feature:	Power Plants
		Stage:	Design
River/Stream:	Elm Fork	Keyword(1):	Carbon
Capacity:	2.2 MW	Keyword(2):	Seals
Nearest City/State:	Lewisville, TX	Keyword(3):	Flexing
Contact Name:	Robert E. Nelson		
Contact Title:	Executive Director of Utilities		
Address:	215 East McKinney		
City/State/Zip:	Denton, TX 76201		
Telephone:	(817) 566-8230		

LESSON LEARNED

Problem: Tyton turbine seal is flexing. This tends to break carbon sealing rings.

Cause & Effect: Turbine hydraulic unit is collecting water from the tyton seal.

Investigations: Turbine manufacturer added a compensation regulator. Also, the new added dump relief valve on the third seal camber of the Tyton seal is very hard to keep adjusted. This valve has very tight set point and constantly goes out of adjustment.

Analysis/Action: The City of Denton is considering taking the seal apart during the next Corps of Engineers outage for inspection. A new plastic valve to replace the compensation regulator, removal of the dump valve from the system, and pulling line 9 from oil reservoir to verify if this arrangement solves the problem, is being considered.

Lessons Learned: Investigate similar sites before starting any design.

Publications: None.

Other Comments: None.

Project:	St. Anthony Falls - Lower Dam	Index No:	PD-007
Owner:	Northern States Power Company	Feature:	Power Plants
		Stage:	Operation
River/Stream:	Mississippi River	Keyword(1):	Sandstone
Capacity:	8.0 MW	Keyword(2):	Foundation
Nearest City/State:	Minneapolis, MN	Keyword(3):	Piping

Contact Name:	Jack J. Schultz
Contact Title:	Administrator, Regulatory Liaison
Address:	919 Nicollet Mall
City/State/Zip:	Minneapolis, MN 55401-1927
Telephone:	(612) 330-5500

LESSON LEARNED

Problem: Powerhouse substructure was undermined collapsing the powerhouse.

Cause & Effect: Piping of erodible sandstone foundation. Rapid increase in flow, undermined limestone foundation.

Investigations: Visual examination of exposed bedrock and collapsed structure.

Analysis/Action: Temporary dike constructed to restore pond. Powerhouse subsequently removed.

Lessons Learned: Erodible foundations require positive cutoff and adequate, monitorable seepage control provisions.

Publications: None.

Other Comments: Modification of adjacent dam led to shortening of seepage path and consequent increase in hydraulic gradient. Tailrace apron drains designed to prevent piping could not be inspected or monitored without dewatering.

Project:	Pit 3 Fishwater Release	Index No:	PC-001
Owner:	Pacific Gas & Electric Co.	Feature:	Power Plants
		Stage:	Construction
River/Stream:	Pit River	Keyword(1):	Fish
Capacity:	NA	Keyword(2):	Welds
Nearest City/State:	Burney , California	Keyword(3):	NDT
Contact Name:	Blake Rothfuss		
Contact Title:	Civil Engineer		
Address:	77 Beale Street F1736A		
City/State/Zip:	San Francisco, CA 94106		
Telephone:	(415) 973-9452		

LESSON LEARNED

Problem: To perform fish studies, an interim fishwater release was required at the Dam regulator. An orifice bulkhead was installed in one of the three sluiceways. Air venting was applied to the downstream face of the bulkhead and the air vent for the sluice gate was capped to prevent gallery flooding.

Cause & Effect: The air vent (sluice gate) cap leaked allowing an air column to form in the water passage. Dynamic loading by the orifice flow was amplified by the air column and the steel bulkhead experienced extensive fatigue at all joints. Inadequate shop inspection allowed the fabrication to use partial penetration butt weld joints instead of full penetration groove welds.

Investigation: The welds and structural members were examined using UT and MT techniques.

Analysis/Action: All welds were field repaired over a six month period. When fatigue failure occurred in the flange to web areas of the structural members, the bulkhead was replaced with a concrete plug with an orifice.

Lessons Learned: Air columns in pressure conduits can complicate structural loading and designs. Improvement in fabrication shop inspection is needed.

Publications: None.

Project:	Holtwood Dam	Index No:	PC-002
Owner:	Pennsylvania Power & Light	Feature:	Power Plants
		Stage:	Construction
River/Stream:	Susquehanna River	Keyword(1):	Movement
Capacity:	102 MW	Keyword(2):	Reactive
Nearest City/State:	Lancaster, PA	Keyword(3):	Aggregate

Contact Name:	G. David Hopfer
Contact Title:	Project Engineer
Address:	2 N. 9th Street, N5
City/State/Zip:	Allentown, PA 18101
Telephone:	(215) 774-6816

LESSON LEARNED

Problem: Over an 80 year period, portions of the powerhouse concrete have heaved and tilted, causing misalignment of hydro turbine-generator equipment and noticeable cracks.

Cause & Effect: Reactive aggregate was used in the original mass concrete construction. No expansion joints were provided.

Investigations: Petrographic analyses of concrete cores confirmed the presence of alkali-aggregate reaction. Annual elevation surveys and crack width measurements were compiled and plotted.

Analysis/Action: Plots showed a slow continuation of substructure movement. Weakened areas were reinforced and hydromachinery was realigned.

Lessons Learned: 1. Avoid reactive aggregate. 2. Provide means of adjusting hydromachinery to accommodate long term substructure movement.

Publications: Hydro Review, Volume V. No. 3, 1986, p. 40.

Other Comments: An attempt was made to seal the interior of wheel pits and scroll cases to cut off the water supply to the alkali aggregate reaction. This was unsuccessful. Reinforcement of weakened areas entailed surface repairs, post-tensioned tendon installation and rock anchor installation. This was considered successful.

Project:	Upriver Dam Hydroelectric Project	Index No:	PO-001
Owner:	City of Spokane, Spokane, WA	Feature:	Power Plants
		Stage:	Operations
River/Stream:	Spokane River	Keyword(1):	Load Rejection
Capacity:	17.7 MW	Keyword(2):	EAP
Nearest City/State:	Spokane, WA	Keyword(3):	Redundancy

Contact Name:	George W. Miller, P.E.
Contact Title:	Design Engineer, Water & Hydroelectric Services
Address:	914 E North Foothills Drive
City/State/Zip:	Spokane, WA 99207-2794
Telephone:	(509) 625-7800

LESSON LEARNED

Problem: A lightning storm caused a full load rejection of all 5 generating units. Back-up power from the emergency generator to operate the gates at the diversion dam could not be restored due to the malfunction of the automatic transfer switch. The city was not able to mobilize a crew quick enough to crank the gates open before overtopping and washout of the embankment occurred.

Cause & Effect: Closure of the turbine wicket gates and failure to open the spillway gates resulted in overtopping the canal embankments. This overtopping damaged the abutments, power canal embankment, powerhouse foundation, and powerhouse equipment. The sudden tripping caused the wicket gate servomotor in one turbine to overtravel resulting in a bent piston rod.

Investigations: An investigation report of the failure was prepared by R. W. Beck and Associates. The Federal Emergency Management Agency's (FEMA) Interagency Hazard Mitigation Team prepared a report to address the causes of this failure and issue directions to appropriate agencies in order for them to take appropriate actions to avert such incidents.

Analysis/Action: Recommendations were: examine the freeboard adequacy in terms of the response time available during a full load rejection; provide redundant auxiliary power for station service and for operating gates; install stops on wicket gate arms to prevent over-travel and provide anti-lift "holder" guides on the operating ring to prevent vertical movement.

Lessons Learned: An Emergency Action Plan (EAP) should address the loss of power scenario; operating personnel must be trained to implement this plan; redundant safety systems must be installed; joining of all engineering disciplines in project design or additions considering a worst-case scenario.

Publications: FEMA Report of November 10, 1986 - FEMA-769-DR City of Spokane, Wa;

Mallur R. Nandagopal, et al: "Hydroelectric Dam washout study guides reconstruction" - Power Engineering, December 1992

Project:	Maxwell Kohler, Sunshine, Oradell & Betasso Power Plants	Index No:	PO-002
Owner:		Feature:	Power Plants
		Stage:	Operations
River/Stream:	Barker Dam & City Water System	Keyword(1):	Gates
Capacity:	3.9 MW (total)	Keyword(2):	Diversion
Nearest City/State:	Boulder, CO	Keyword(3):	Generators
Contact Name:	John Cowdrey		
Contact Title:	Hydroelectric Technician		
Address:	1094 Betasso Road		
City/State/Zip:	Boulder, CO 80303		
Telephone:	(303)441-3245		

LESSON LEARNED

Problem: Hydroelectric plants were installed in-line to an existing water treatment system. During operation of the power plants the following problems were experienced: power plants were not performing to expected level of generation, control problems caused generator trips and unreliable operation, capacity tests caused sand to be introduced into the treated water.

Cause & Effect: Generator trips were caused by control and equipment problems, which were investigated and resolved. In two cases the control system was not suited for the application and proved to be unreliable. Sand that had settled into the clearwell was being sucked into the water system during required monthly capacity tests. Existing undersized distribution piping caused back pressure and head loss, resulting in limited generator performance. Opening of bypass valves during peak water demand periods further aggravated the low-power, head loss and back pressure problems. Also, one generator could not provide rated power due to overheating. One bypass valve was a ball valve, which was not suited to the application.

Investigations: A computer model of the piping system flow is being done to determine where the piping was undersized, and if it can be replaced. An analysis of the overheating generator determined that the machine was not designed correctly. The City hired operation and maintenance specialists to troubleshoot, maintain and operate the hydro system.

Analysis/Action: Undersized piping will be replaced where possible. Experienced maintenance personnel were hired. Two control systems were replaced with Programmable Logic Controllers. Procedures were changed when the capacity tests were being performed. The overheating generator will be replaced. The ball valve was replaced with a non-cavitating design sleeve valve.

Lessons Learned: Hydro power plants which operate in-line to a water treatment and distribution systems need to be designed for all distribution flow scenarios, including peak flows and projected future flows. Operators of hydro power plants which are to be operated in non-traditional flow patterns such as water distribution systems need special training. The design of a hydro power plant which operates with water supply systems requires engineers with previous experience in such designs. Water treatment personnel must be involved in the hydro plant and control system design to ensure that the hydro plant meets their unique water distribution needs.

Publications: None.

Project:	Lloyd Shoals Hydro Plant	Index No:	PO-003
Owner:	Georgia Power Company	Feature:	Power Plants
		Stage:	Operation
River/Stream:	Ocmulgee, Jackson Lake	Keyword(1):	Flashboards
Capacity:	4.4 MW	Keyword(2):	Corrosion
Nearest City/State:	Jackson, Georgia	Keyword(3):	Leakage
Contact Name:	J. R. Pope		
Contact Title:	Plant Manager		
Address:	P.O. Box 3640		
City/State/Zip:	Eatonton, Georgia 31024		
Telephone:	(706) 485-2000 or (404) 526-3608		

LESSON LEARNED

Problem: Difficulty in resetting dam flashboards.

Cause & Effect: The flashboards had corroded and were leaking excessively. The condition of the flash boards also made it difficult to reset after tripping.

Investigation: It was found that corrosion had reduced the weight of the flashboard counter weights and was causing the difficulty in resetting.

Analysis/Action: A project to strengthen the dam required removal of the flashboard system. After this project was complete all of the steel flashboard components were replaced with galvanized steel and all of the wood components were replaced with pressure treated creosote boards. The new steel components will be monitored to evaluate the condition of the galvanized steel when emersed for long periods.

Lessons Learned: The new flashboards significantly reduced leakage and were easier to reset. Suggest researching automatic reset for flashboards.

Publications: None.

Project:	Cuero	Index No:	PO-004
Owner:	Cureo Hydroelectric Partnership	Feature:	Power Plants
		Stage:	Operation
River/Stream:	Guadalupe River	Keyword(1):	Foundation
Capacity:	1.047 MW	Keyword(2):	Settlement
Nearest City/State:	Cuero, Tx	Keyword(3):	Piping
Contact Name:	John L. McNeill		
Contact Title:			
Address:	P.O. Box 2211		
City/State/Zip:	Victoria, Texas 77902		
Telephone:	(512) 578-3032		

LESSON LEARNED

Problem: Settlement and distortion of powerhouse foundation

Cause & Effect: After the old powerhouse was rehabilitated the forebay was dredged. This resulted in the settlement of the powerhouse. This settlement was caused by ground water seeping piped soil from under the powerhouse.

Investigation: No official investigation was done, but some borings by local drillers provided a rough idea of the subsurface conditions and indicated the sandstone foundation to be partially cemented.

Analysis/Action: The foundation was re-leveled, a cutoff wall was installed. After the cut-off wall was installed the powerhouse was re-leveled and damage was repaired.

Lessons Learned: Lack of geologic information under a powerhouse can result in damage to the structure. A detailed geotechnical investigation is necessary when constructing a powerhouse or making modifications.

Publications: None.

Project:	Toledo Bend Project Joint Operation	Index No:	PO-005
Owner:	Sabine River Authority of Texas and	Feature:	Power Plants
	Sabine River Authority of Louisiana	Stage:	Operation
River/Stream:	Sabine River	Keyword(1):	Erosion
Capacity:	81 MW	Keyword(2):	Draft Tube
Nearest City/State:	Burkeville, Texas, Anacoco, Louisiana	Keyword(3):	Lining

Contact Name:	Bartom Rumsey
Contact Title:	Project Engineer
Address:	Rt. 1, Box 780
City/State/Zip:	Many, Louisiana 71449-9730
Telephone:	(318) 256-4112

LESSON LEARNED

Problem: Erosion was found in the draft tube steel liner immediately above and below the runner blade tips as well as on the bottom trailing side of each runner blade.

Cause & Effect: The erosion on the liner and blades had caused a loss of unit efficiency.

Investigation: Vibration tests and efficiency indexing tests were performed on each unit.

Analysis/Action: The erosion was repaired with a welded stainless steel overlay. After the repairs the units were prepared for comparison tests.

Lessons Learned:

Publications: None.

Project:	Cherokee Falls Hydroelectric Facility	Index No:	PO-006
Owner:	Broad River Electric Cooperative	Feature:	Power Plants
		Stage:	Operation

River/Stream:	Broad River	Keyword(1):	Flooding
Capacity:	82.447 MW	Keyword(2):	Flashboards
Nearest City/State:	Gaffney, S. C.	Keyword(3):	Debris

| Contact Name: | R. C. Carroll and E. D. Metivier |
| Contact Title: | |

Address:	P.O. Box 790
City/State/Zip:	Gaffney, SC 29340
Telephone:	(803) 329-0644

LESSON LEARNED

Problem: High river flows (65,800 cfs) over flooded the powerhouse and private property, damaging equipment and left bank of the Broad River.

Cause & Effect: The flooding was caused by accumulation of trees and debris on 1700 lineal feet of the flashboards and inadequate flood protection along the left river bank. The elevation of the river bank was too low in comparison to flood elevations.

Investigation: Separate investigation by FERC staff.

Analysis/Action: The FERC required additional analysis of the flooding to determine the mitigation measures required and submission of an acceptable Emergency Action Plan before the powerhouse could be put back in to operation.

Lessons Learned: The design of the flashboards and trash rakes need to consider accumulation of debris. The flood of record is not necessarily the maximum design flood to be experienced by a newly installed facility.

Publications: DOE/ID/12125-1,2 (National Technical Information Service) "Small Scale Hydro Program- Feasibility Assessment and Technology development-Summary Report"

Project:	Ellis Hydroelectric Plant	Index No:	PO-007
Owner:	Arkansas Electric Coop. Corp.	Feature:	Power Plants
		Stage:	Operation
River/Stream:	Arkansas River	Keyword(1):	Corrosion
Capacity:	33 MW	Keyword(2):	Stainless Steel
Nearest City/State:	Fort Smith, AR	Keyword(3):	Pitting

Contact Name:	Maurice Robinson
Contact Title:	Director - Engineering, Construction & Operations
Address:	P.O. Box 194208
City/State/Zip:	Little Rock, AR 72219-4208
Telephone:	(501) 570-2497

LESSON LEARNED

Problem: Water in the Arkansas River is corrosive to stainless steel discharge rings. Numerous small pits were noted in each ring after the plant was in operation.

Cause & Effect: The corrosive river water caused pits in the discharge rings.

Investigations: The river water was analyzed and past studies reviewed. The pitting of the metal was analyzed.

Analysis/Action: The pitting was repaired and cathodic protection will be installed.

Lessons Learned: The Arkansas River water has a greater effect on this stainless steel alloy than other river waters have. Another similar plant is under construction and this lesson will help prevent problems there.

Publications: None.

Other Comments: Stainless steel may not be the best corrosion resistant material to a fresh water exposure.

Project:	Colebrook Hydro Project	Index No:	PO-008
Owner:	MDC	Feature:	Power Plants
		Stage:	Operation
River/Stream:	W. Branch Farmington	Keyword(1):	Computer
Capacity:	3 MW	Keyword(2):	Virus
Nearest City/State:	Colebrook, CT	Keyword(3):	Software

Contact Name:	Miles Messenger
Contact Title:	Hydro Superintendent/Manager of Water Supply
Address:	39 Beach Rock Road
City/State/Zip:	Barkhamsted, CT 06063
Telephone:	(203) 379-0938

LESSON LEARNED

Problem: Operational computer acquired destructive "virus" in hard drive. All backups of software also were infected.

Cause & Effect: Virus introduced through exchange/upgrade of software from programmer.

Investigations: Acquired virus detection software and found resident in operational files and hard drive of computer.

Analysis/Action: Contacted vendor to reformat hard drive (erase) and installed resident virus protection software. All software installed was then run against two different virus detection programs and eradicated prior to installation.

Lessons Learned: Do not install software before it has been checked for viruses especially if not from a commercial source or if not new.

Publications: None.

Other Comments: None.

Project:	Cougar Dam	Index No:	PO-009
Owner:	U.S. Army Corps of Engineers	Feature:	Power Plants
		Stage:	Operation
River/Stream:	South Fork of McKenzie River	Keyword(1):	Wicket Gate
		Keyword(2):	Galling
Capacity:	2 Units, 12.5 MW each	Keyword(3):	Spatter
Nearest City/State:	Blue River, OR		
Contact Name:	Rodney J. Wittinger, P.E.		
Contact Title:	Mechanical Engineer		
Address:	P.O. Box 2870, CENPD-PE-HD		
City/State/Zip:	Portland, OR 97208		
Telephone:	(503) 326-3840		

LESSON LEARNED

Problem: Galling of wicket gate, and the top and bottom plates. This problem is a general problem on many turbines Mr. Wittinger has inspected over the last 15 years.

Cause & Effect: Welding to repair turbine runner cavitation results in "metal spatter" on the top and bottom plates. If the "weld spatter" is not removed and the wicket gates are operated, the sealing surfaces of the wicket gates and top and bottom plates are damaged by galling. "Spatter" adherence to the vertical sealing surface of wicket gates, preventing full closure or causing a rough surface and subsequent erosion damage, also has been observed.

Investigations: This problem is reoccurring for both Francis and Kaplan type units. When galling is reported, this is the usual cause according to Mr. Wittinger.

Analysis/Action: Inspect wicket gates and top and bottom plates and remove "spatter" prior to returning the unit to service.

Lessons Learned: This has been a general problem on any unit which receives weld repair. Many turbine inspections have indicated this is not an isolated problem. A detailed inspection of the areas subject to damage after all welding repair work is complete is required. Remove the "weld spatter" by chipping or grinding. Take care that proper tools are used. This will prevent a corrosion cell which can cause an equivalent problem.

Publications: Unknown

Other Comments: Assure that the maintenance staff has adequate time to perform a complete job, and that the inspection is included in the "Punch List" prior to returning the unit to service.

Project:	Curecanti Morrow Point Dam	Index No:	PO-010
Owner:	US Bureau of Reclamation	Feature:	Power Plants
		Stage:	Operation
River/Stream:	Gunnison River	Keyword(1):	Pump
Capacity:	120 MW (2-60 MW units)	Keyword(2):	Mineral
Nearest City/State:	Montrose, CO	Keyword(3):	Deposits
Contact Name:	David M. McAfee		
Contact Title:	Chief, Power & Storage Division		
Address:	HC-30, Box 312		
City/State/Zip:	Truth or Consequences, NM 87901		
Telephone:	(505) 894-6661		

LESSON LEARNED

Problem: A mineral deposit from the water would build up on the pump impellers and bowls of the vertical turbine pumps.

Cause & Effect: The mineral deposit would build up to the point that it would freeze the pump.

Investigations: Throttled pump down and ran continuously helped, but it was hard to match inflow and standby pump built up.

Analysis/Action: Samples of deposit and water with pump materials were sent to lab for analysis. A drip water softener was installed in sump. Don't remember chemical.

Lessons Learned: Water analysis of water should be made before specifying pumps so pump materials could be specified that would not react to the water pumped.

Publications: None.

Other Comments: None.

Project:	Glen Canyon Dam	Index No:	PO-011
Owner:	US Bureau of Reclamation	Feature:	Power Plants
		Stage:	Operation
River/Stream:	Colorado River	Keyword(1):	HVAC
Capacity:	1340 MW	Keyword(2):	Access
Nearest City/State:	Page, AZ	Keyword(3):	Platforms

Contact Name:	Gary Kuhstoss
Contact Title:	Chief, Glen Canyon Field Branch
Address:	P.O. Box 1477
City/State/Zip:	Page, AZ 86040
Telephone:	(602) 645-2481

LESSON LEARNED

Problem: There was no access provided to heating and ventilating units in the generator room.

Cause & Effect: Ladders had to be used to change filters.

Investigations: Consideration was given to using a portable hydraulic lift or constructing access platforms.

Analysis/Action: Access platforms were locally designed, fabricated and installed.

Lessons Learned: Walking and working surfaces should be provided to minimize safety risk to employees.

Publications: None.

Other Comments: This problem was primarily one of employee safety. In this case, employees were observed walking a narrow ledge for access to the heating and ventilating units rather than finding and using a ladder or portable lift.

Project:	Glen Canyon Dam	Index No:	PO-012
Owner:	US Bureau of Reclamation	Feature:	Power Plants
		Stage:	Operation
River/Stream:	Colorado River	Keyword(1):	Generator
Capacity:	1,340 MW	Keyword(2):	Rotor
Nearest City/State:	Page, AZ	Keyword(3):	Inspection
Contact Name:	Gary Kuhstoss		
Contact Title:	Chief, Glen Canyon Field Branch		
Address:	P. O. Box 1477		
City/State/Zip:	Page, AZ 86040		
Telephone:	(602) 645-2481		

LESSON LEARNED

Problem: Access to area beneath rotor of generator.

Cause & Effect: Access to the area under the rotor is by crawling under the stator or by opening a bolted hatch cover at the top of the ladder from the turbine pit.

Investigations: Walkway to this area could have been provided to eliminate crawling through the narrow space underneath the stator.

Analysis/Action: No action has been taken to alleviate the problem.

Lessons Learned: Adequate access should be provided to areas where work must be done.

Publications: None.

Other Comments: Access is needed to the area under the rotor so that inspections of the lower end turns of the stator, the lower part of rotor and the lower guide bearing, rotation detector, runout sensors, brakes, and other equipment can be inspected and maintained.

Project:	Glen Canyon Dam	Index No:	PO-013
Owner:	US Bureau of Reclamation	Feature:	Power Plants
		Stage:	Operation
River/Stream:	Colorado River	Keyword(1):	Valve
Capacity:	1340 MW	Keyword(2):	Access
Nearest City/State:	Page, AZ	Keyword(3):	Platforms
Contact Name:	Gary Kuhstoss		
Contact Title:	Chief, Glen Canyon Field Branch		
Address:	P. O. Box 1477		
City/State/Zip:	Page, AZ 86040		
Telephone:	(602) 645-2481		

LESSON LEARNED

Problem: Difficulty in reaching 12-inch generator cooling water valves.

Cause & Effect: No access platform was provided to reach these valves so they were originally operated from a ladder.

Investigations:

Analysis/Action: Access platforms were locally designed, fabricated and installed to improve access to these valves.

Lessons Learned: Walking and working surfaces should be provided to valves and other equipment to minimize risk to employees.

Publications: None.

Other Comments: These valves were not the only pieces of equipment needing better access.

Project:	Holtwood Dam	Index No:	PO-014
Owner:	Pennsylvania Power & Light	Feature:	Power Plants
		Stage:	Operation
River/Stream:	Susquehanna	Keyword(1):	Random
Capacity:	102 MW	Keyword(2):	Vibration
Nearest City/State:	Lancaster, PA	Keyword(3):	Vortex
			Shedding
Contact Name:	G. David Hopfer		
Contact Title:	Project Engineer		
Address:	2 N. 9th Street, N5		
City/State/Zip:	Allentown, PA 18101		
Telephone:	(215) 774-6816		

LESSON LEARNED

Problem: Occurrences of random vibration, up to 50 mils in amplitude, were measured at the lower turbine guide bearing shortly after the initial start up of a renovated double runner vertical Francis turbine.

Cause & Effect: Cause of the vibration was unknown; although due to the randomness, it was suspected to be related to draft tube hydraulics. Vibration was present only at 90%-100% gate opening. Concerns were about wear on the guide bearing.

Investigations: The lower turbine shaft rotates submerged in the upper draft tube. Vortex shedding frequencies and drag forces were computed and compared to operating frequencies.

Analysis/Action: "Anti-vortex rings" were fabricated and clamped to the submerged portion of the shaft.

Lessons Learned: Following installation of the anti-vortex rings, lower turbine guide bearing vibrations dropped to 8 mils and remained steady. Power tests showed no change in output. The solution was deemed a success.

Publications: None.

Other Comments: The lower guide bearings are water lubricated and have hard rubber staves. The other non-renovated units in the powerhouse are not equipped with vibration probes at the lower guide bearing, so it is unknown whether the vibration is attributable to the original 83 year old design, or a result of renovation.

Project:	Holtwood Dam	Index No:	PO-015
Owner:	Pennsylvania Power & Light	Feature:	Power Plants
		Stage:	Operation
River/Stream:	Susquehanna	Keyword(1):	Dewatering
Capacity:	102 MW	Keyword(2):	Generators
Nearest City/State:	Lancaster, PA	Keyword(3):	Wheel Pit
Contact Name:	G. David Hopfer		
Contact Title:	Project Engineer		
Address:	2 N. 9th Street, N5		
City/State/Zip:	Allentown, PA 18101		
Telephone:	(215) 774-6816		

LESSON LEARNED

Problem: The wheel pit dewatering system (pumps, headers and valves) became unreliable due to leaks and pump problems.

Cause & Effect: The system was in need of complete overhaul. Maintenance outages were being delayed up to two weeks due to dewatering system problems and other leakage.

Investigations: Submersible pumps were proposed as an alternative to the 80 year old dewatering system.

Analysis/Action: Submersible pumps were bolted to the water side of draft tube stoplogs. Pumps take suction through penetrations in the stoplogs and discharge through a check valve directly to the tailrace.

Lessons Learned: This system proved very reliable and was an inexpensive alternative to dewatering system overhaul.

Publications: Hydro Review, June 1993, pp. 12-14.

Other Comments: Pumps operate on float controls. Check valves are simple "flat plate flaps". Plant operators, mechanics and construction personnel performed most of the design work, under review by engineering staff.

Project:	Hoover Dam	Index No:	PO-016
Owner:	US Bureau of Reclamation	Feature:	Power Plants
		Stage:	Operation
River/Stream:	Colorado River	Keyword(1):	Generators
Capacity:	1,935 MW	Keyword(2):	Controls
Nearest City/State:	Boulder City, NV	Keyword(3):	Air Bubble
Contact Name:	James D. Sloan		
Contact Title:	General Engineer		
Address:	P.O. Box 25007, Denver Federal Center		
City/State/Zip:	Denver, CO 80225		
Telephone:	(303) 236-8085		

LESSON LEARNED

Problem: The air supply did not maintain the air bubble for the units under all conditions when the units were on line for spinning reserve. Loss of the bubble dramatically increases the power consumption of a unit.

Cause & Effect: Pressure switches were used to maintain sufficient head cover air pressure to force the draft tube water level below the wheel, however, different tailrace levels required a pressure adjustment to work properly.

Investigations: After the investigation, it was suggested to monitor power consumption to control the air admission system. A circuit was built to detect output from a megawatt transducer and was tested successfully on one unit.

Analysis/Action: The megawatt detectors were set about 5% greater than normal motoring power consumption, they operate the air valves for five seconds beyond the time required to reduce unit power consumption below the detector setting.

Lessons Learned: The use of unit power consumption to control air admission for motoring is both practical and inexpensive. It requires no complex control schemes and no equipment in the draft tube.

Publications: Bureau of Reclamation Power Operations and Maintenance Workshop, October 20-22, 1992, Hydro Review, April 1992 (Vol. XI, No. 2), p. 100; Sticky Wickets: "New Approach to Controlling Air Admission During Monitoring."

Other Comments: None.

Project:	Lakeport Hydroelectric Project	Index No:	PO-017
Owner:	NH Water Resources Council/Lakeport Hydro Corp.	Feature:	Power Plants
		Stage:	Operation
River/Stream:	Winnepesaukee River	Keyword(1):	Power Sales
Capacity:	0.705 MW	Keyword(2):	Automation
Nearest City/State:	Laconia, NH	Keyword(3):	Controls
Contact Name:	Irv Toiles		
Contact Title:	President		
Address:	P.O. Box 240		
City/State/Zip:	Manchester, NH 03105		
Telephone:	(603) 669-3822		

LESSON LEARNED

Problem: Overall WATER-TO-WIRE efficiency needed to be improved -- specifically TIME-OF-DAY operations.

Cause & Effect: Unmanned, manual station operation could not take full advantage of the TIME-OF-DAY power sales contract. Power production and revenue suffered.

Investigations: A cost-benefit study was conducted to determine the economic feasibility of installing automated station controls.

Analysis/Action: New, fully automated station operation controls (with touch tone remote control capability) were installed. Power production and revenues were improved as well as reservoir management.

Lessons Learned: Higher up-front capital costs are off set by improved long-term power production and lower personnel/operational costs.

Publications: None.

Other Comments: None.

Project:	Libby Dam	Index No:	PO-018
Owner:	U.S. Army Corps of Engineers	Feature:	Power Plants
		Stage:	Operation
River/Stream:	Kootenai River	Keyword(1):	Runout
Capacity:	525 MW	Keyword(2):	Vibration
Nearest City/State:	Libby, MT	Keyword(3):	Generator
Contact Name:	Bob Schloss		
Contact Title:	Project Manager		
Address:	17115 Highway #37		
City/State/Zip:	Libby, MT 59923-9703		
Telephone:	(406) 293-7751		

LESSON LEARNED

Problem: Rough operation (vibration and noise) of Unit 4.

Cause & Effect: One shear pin broke during a routine shut down. After the shear ----- was replaced the unit was restarted 1-2 days later and ran rough and noisy (particularly at best efficiency).

Investigations: Unit was unwatered and inspected by maintenance forces on December 1, 1992. Nothing found. Preliminary tests were performed week of December 21. During tests, a broken shear pin was found.

Analysis/Action: Replaced shear pin.

Lessons Learned: This was a simple failure but hard to see. All wicket gates moved when the unit was unwatered. Gate levers hide shear levers when Unit is stopped.

Publications: None.

Other Comments: It is very important that the obvious cause of the problem be investigated. The Unit was shut down for three weeks until engineers could inspect/test it. The problem was immediately obvious (rough operation) after a short routine shutdown.

Project:	Ocoee III Hydro Plant	Index No:	PO-019
Owner:	Tennessee Valley Authority	Feature:	Power Plants
		Stage:	Operation
River/Stream:	Ocoee River	Keyword(1):	Thrust Bearing
Capacity:	28.8	Keyword(2):	Sole Plates
Nearest City/State:	Copper Hill, TN	Keyword(3):	Foundation Bolts

Contact Name:	Michael A. Thompson
Contact Title:	Hydro Specialist
Address:	1101 Market Street, BR 3D-C
City/State/Zip:	Chattanooga, TN 37402
Telephone:	(615) 751-3043

LESSON LEARNED

Problem: The unit thrust bearing system was reworked frequently. The runner plate, spacer, and the generator shaft exhibited signs of excessive movement; this resulted in fretting and loose fit of the components.

Cause & Effect: The sole plates for the thrust bridge were found to be moving. This movement was allowed by loose anchor bolts. The continuous movement had powdered the grout around the anchor bolts.

Investigations: Indicators were attached to each bridge leg and then monitored. The machine was taken through a full range of start-up, condensing, generating, and shut down operations to observe movements.

Analysis/Action: The generator was disassembled and the sole plates removed. The foundation bolts were loose. The bolts were core-drilled to allow regrouting. The sole plates were reset and leveled.

Lessons Learned: Look for structural defects and loose foundation bases that may allow movement of sole plates.

Publications: None.

Other Comments: Proper torque should be applied to foundation bolts. Foundation bolts should be checked on a periodic basis.

Project:	Ray Roberts Dam	Index No:	PO-020
Owner:	City of Denton, TX	Feature:	Power Plants
		Stage:	Operation
River/Stream:	Elm Fork	Keyword(1):	Equalizer
Capacity:	1.2 MW	Keyword(2):	Turbine
Nearest City/State:	Aubrey, TX	Keyword(3):	Orifice
Contact Name:	Robert E. Nelson		
Contact Title:	Executive Director of Utilities		
Address:	215 East McKinney		
City/State/Zip:	Denton, TX 76201		
Telephone:	(817) 566-8230		

LESSON LEARNED

Problem: Orifice plate adjustments for turbine equalizer. Replace lube oil flow switch and water flow control valve. Generator speed switch adjustment for over and underspeed protection. The control board for excitation system is sensitive to power fluctuations.

Cause & Effect: Orifice plate adjustment to prevent water leakage on the powerhouse floor. Lube oil flow switch and water flow control valve will be replaced due to startup problem.

Investigation: None.

Analysis/Action: Drill more holes in orifice plates. Adjust flow set points in correlation with oil temperature in lube oil system. Speed switch goes out of calibration after a period of operation.

Lessons Learned: Investigate similar sites before starting any design.

Publications: None.

Other Comments: Verify commissioning procedures in advance.

Project:	Roanoke Rapids/Gaston Hydro Station	Index No:	PO-021
		Feature:	Power Plants
Owner:	Virginia Electric & Power Co.	Stage:	Operation
River/Stream:	Roanoke River	Keyword(1):	Cavitation
Capacity:	104 MW/225 MW	Keyword(2):	Overlays
Nearest City/State:	Roanoke Rapids, NC	Keyword(3):	Stainless Steel

Contact Name:	Clarence H. Powell
Contact Title:	Operating Supervisor, Roanoke Rapids Hydro
Address:	P.O. Box 370
City/State/Zip:	Roanoke Rapids, NC 27870
Telephone:	(919) 535-6166

LESSON LEARNED

Problem: Earlier methods of performing cavitation repairs using composite or ceramic coatings by several different suppliers did not prove successful for the long term and made subsequent repairs more costly and time consuming.

Cause & Effect: Various coatings had been used; all these coatings saved time initially but did not hold up. Removing these coatings which were damaged by subsequent cavitation was difficult and required extensive grinding since burning would not remove it.

Investigations: Subsequent repairs by weld-overlay required labor intensive time to prepare.

Analysis/Action: Currently all cavitation repairs are performed by grinding and arc-gouging. Weld overlay build-up using a stainless steel 309L rod.

Lessons Learned: Cavitation repairs made using SS 309L are more resistant to cavitation damage and thus provide greater protection and longer times between necessary cavitation outages.

Publications: None.

Other Comments: Replacement turbine runners (Kaplan type) with CA-6NM stainless steel blades provides much improved cavitation resistance as compared to cast carbon steel blades.

Project:	Dos Amigos Pumping Plant	Index No:	PO-022
Owner:	State of California	Feature:	Power Plants
	Department of Water Resources	Stage:	Operation
River/Stream:	California Aqueduct	Keyword(1):	Wear Rings
Capacity:	15,450 cfs	Keyword(2):	Pumps
Nearest City/State:	Santa Nella, CA	Keyword(3):	
Contact Name:	Paul Flanagan		
Contact Title:	Chief of Engineering		
Address:	31770 W. Highway 152		
City/State/Zip:	Santa Nella, CA 95322		
Telephone:	(209) 826-0718		

LESSON LEARNED

Problem: Retaining bolts in the stationary wearing ring had broken loose and were being ground up against the impeller.

Cause & Effect: The wear ring was being hammered on shut down. Since the units run in reverse overspeed causing vibration, this caused the bolts to back out and get broken, and then be ground between the wear ring and the impeller.

Investigations: A 3/8-inch hole was drilled through the pump casing then a borescope was inserted to see if the hold down bolts were being broken. The rotor was turned so that each bolt would appear opposite the hole.

Analysis/Action: First, the impeller was raised with the oil lift pump. Then a come-along was attached and the impeller turned. Wherever there should be a retaining bolt, they looked through a 3/8-inch observation hole which was drilled through the impeller skirt. They were able to inspect all 36 of the ring retaining bolts: 21 of them had broken and parts of some were missing. They enlarged the 3/8-inch hole to two inches. This size hole gave them room to replace broken bolts and to weld their heads. To do this, they jacked up the impeller and moved it around the circle once more. Each broken bolt was replaced and tack welded to prevent backing out. The wearing ring was secured in place (temporarily), then the holes in the impeller skirt were plugged and welded closed.

Lessons Learned: To check these bolts often.

Publications: State of California, Department of Water Resources, Technical Bulletin Number 89, April through September 1992.

Other Comments: None.

Project:	Dos Amigos Pumping Plant	Index No:	PO-023
Owner:	State of California	Feature:	Power Plants
	Department of Water Resources	Stage:	Operation
River/Stream:	California Aqueduct	Keyword(1):	Rotors
Capacity:	15,540 cfs	Keyword(2):	Motors
Nearest City/State:	Santa Nella, CA	Keyword(3):	Laminations
Contact Name:	Paul Flanagan		
Contact Title:	Chief of Engineering		
Address:	31770 W. Highway 152		
City/State/Zip:	Santa Nella, CA 95322		
Telephone:	(209) 826-0718		

LESSON LEARNED

Problem: To disassemble, clean and restack 10,400 steel laminations that make up a 100 ton, 18-foot diameter rotor for a 40,000 HP pump motor.

Cause & Effect: The nuts which secure the laminations had twisted the 5-foot long bolts which held the laminations together. Later the stresses caused by heat, vibration and reverse forces on the pump caused the bolts to relax and allow the laminations to move.

Investigations: Investigation revealed that the poles were larger in diameter and had run the hottest at the mid height position.

Analysis/Action: After tear-down, the immediate problem was to clean each side of the laminations. A pair of the round brushes used by an ordinary, rotary floor buffer was used for the scrubbing part. The brushes were set up facing one another, attached by a hinge to a framework. With this system, a lamination could be inserted between the two rotating brushes which did the scrubbing. To give control and support to the rather flexible laminations, a guide trough made of angle iron was provided. The laminations could now be pushed (and pulled) along this trough, between the brushes, and so on down the line. Rinsing followed the brushing; pipes with nozzle-like openings aimed sprayed on the laminations. Beyond the rinsing device, there was a pair of power blowers which dried the laminations on both sides. The field division used an optical scanner to true up the rim laminations. The optical scanner works in conjunction with a magnetic target to give a digital reading. The workers using the scanner make constant checks to be sure that manufacturer's recommended tolerances are not exceeded. Adjustments can be made by using wedges between the rotor, spider and rim. The stack may also be moved manually with a mallet. The long, doubled-ended bolts that hold the rotor rim iron were replaced by bolts of much stronger steel.

Lessons Learned: New keys and stronger bolts are needed. The bolts were tightened to achieve a stretch of 0.030": each pole had two bolts 60" and 63.5" length.

Publications: State of California, Department of Water Resources, Technical Bulletin, Number 87, July through December 1991.

Other Comments: None.

Project:	B. F. Sisk-San Luis Dam	Index No:	PO-024
Owner:	State of California	Feature:	Power Plants
	Department of Water Resources	Stage:	Operation
River/Stream:	Off Stream Storage	Keyword(1):	Amortissuer
Capacity:	440 MW	Keyword(2):	Windings
Nearest City/State:	Santa Nella, CA	Keyword(3):	Motors
Contact Name:	Paul Flanagan		
Contact Title:	Chief of Engineering		
Address:	31770 W. Highway 152		
City/State/Zip:	Santa Nella, CA 95322		
Telephone:	(209) 826-0718		

LESSON LEARNED

Problem: The equipment and facilities of the project have been in constant use for twenty-seven years. Wear and tear is inevitable; failures and breakdowns are bound to occur.

Cause & Effect: The amortissuer connectors had suffered from high temperature over the years. The layers of metal had begun to spread apart and flake away -- eventually separating completely. This resulted in high-current arcing at the connections.

Investigations: An inspection was made to watch wear and tear of the units when they were started. Deterioration was found in the amortissuer winding connectors on the rotor.

Analysis/Action: Beginning three years ago, each pre-start inspection showed deterioration in the amortissuer winding connectors on the rotor. These connectors, made of laminated copper, had suffered from high temperature over the years. The layers of metal had begin to spread apart and flake away -- eventually separating completely. This resulted in high current arcing at the connections.

Original specifications called for units to be started unwatered. Water was to be introduced after the machine was running. Without the great weight of water and with reduced voltage, a Unit's starting current was relatively low and electrical damage tended to be minimal. It was a good plan. But, in practice, starting unwatered and then introducing water caused too much pressure on the discharge line, so the Units have been started at full voltage and fully watered. This led to the high current problems and deterioration.

Although the entire starting winding takes a beating during start-up, the bar connectors seem to suffer the most. They are the weakest link in the starting winding and they get hottest during starts.

Lessons Learned: Inspect amortissuer windings frequently.

Publications: State of California, Department of Water Resources, Technical Bulletin, Number 76, April, May, June 1988.

Other Comments: None.

Project:	Dos Amigos Pumping Plant	Index No:	PO-025
Owner:	State of California	Feature:	Power Plants
	Department of Water Resources	Stage:	Operation
River/Stream:	California Aqueduct	Keyword(1):	Keys
Capacity:	15,450 cfs	Keyword(2):	Motors
Nearest City/State:	Santa Nella, CA	Keyword(3):	Rotors

Contact Name:	Paul Flanagan
Contact Title:	Chief of Engineering
Address:	31770 W. Highway 152
City/State/Zip:	Santa Nella, CA 95322
Telephone:	(209) 826-0718

LESSON LEARNED

Problem: 40,000 HP rotors are attached to a shaft with tapered keys approximately five feet long. When new thicker keys are to be constructed, how do you get the correct taper? The old keys could not be used as a pattern since the impeller had worked back and forth on the shaft destroying the old keys.

Cause & Effect: When pumping water, the motor turns CCW; when the pump is shut down, the column of water 18-feet in diameter and 1/2 mile long runs back at the pump causing it to stop and turn in a CW direction; this reversal had loosened the motor spider on the shaft.

Investigations: Scribe marks on the shaft and spider indicated movement between them.

Analysis/Action: Repair activity consisted of sending the shaft and hub out to be built up with weld. Then having the mating surfaces machined to insure a good shrink fit. When the spider was expanded, the shaft was lifted and then slid into place. The shaft, hub and spider were secured together with three steel keys in keyway slots. The keys were about three feet long and slightly tapered from top to bottom. The shaft was in proper position when each key slipped snugly into place. However, the spider, hub and shaft had been pulled apart and remachined. The chances were small that the old keys would fit, so new keys of a larger cross sectional area were made. There was no way to be sure of the taper of the slots. The solution was simple. A rectangular ruler-like bar (approximately the length and width of the keyway) was produced. The bar was thinner than a key. At ten-inch intervals, half-inch holes were bored through the bar. Three-quarter-inch pins of steel were installed into each hole. A "C" spring acting between the bore and the pin assured this fit. The tops of the pins were beveled a little to give a ball-bearing-like contact surface. With the pins pushed out as far as possible, the bar was shoved into the keyway slot until it bottomed out. All the pins were extended to their maximum before the bar was placed in the slot. When the bar was shoved into the slot fully, each pin with the "C" spring on its side would adjust to the width of the tapered slot at its location. The bar can be placed on its side, removed, and a line plotted touching the top of each of the pins. When this line is transferred to a drawing of the particular key, it is easy to see what has to be done to produce the correct taper-fit for that individual keyway.

Lessons Learned: Movement between the rotor spider and the shaft should be monitored.

Publications: State of California, Department of Water Resources, Technical Bulletin, Number 84, October, November, December 1990.

Project:	B. F. Sisk-San Luis Dam	Index No:	PO-026
Owner:	State of California	Feature:	Power Plants
	Department of Water Resources	Stage:	Operation
River/Stream:	Off Stream Storage	Keyword(1):	Welder
Capacity:	440 MW	Keyword(2):	Windings
Nearest City/State:	Santa Nella, CA	Keyword(3):	Motor

Contact Name:	Paul Flanagan
Contact Title:	Chief of Engineering
Address:	31770 W. Highway 152
City/State/Zip:	Santa Nella, CA 95322
Telephone:	(209) 826-0718

LESSON LEARNED

Problem: Amortissuer straps are made of laminated copper and are attached to the bars and the adjacent poles by silver solder normally heated with an acetylene torch. Overheating and heating too much of the copper area reduces the life of the strap.

Cause & Effect: When soldering with a torch, the copper is heated above 1200ºF for four or five inches on each side of the soldered joint. Heating copper above 1200ºF reduces the life of the copper by 50%.

Investigations: The investigation confirmed that soldering the heat must be confined to just the soldered area.

Analysis/Action: The resistance welder at Gianelli Pumping-Generating Plant is of Swedish design and manufacture. It is a 12,000-Hertz, high frequency, electrical induction welder. The welding tip is a hollow water-cooled, brass tube. In addition to supplying the high frequency current, the equipment pumped cooling water at the rate of 12 gallons a minute to cool the tip. The manufacturer's tip heated one side of the strap only. This caused overheating of the copper by the time the strap was welded. That meant a loss of copper life. Things would be better with a welding tip that wrapped around the connector strap. Bending the rectangular copper tubing was a tricky chore; however, the crew designed and produced the wraparound tip that was needed. The San Luis crew then experimented with various flux coatings until they found one that worked. The success of this project has been spectacular.

Lessons Learned: A high frequency welder will extend the life of the material many times.

Publications: State of California, Department of Water Resources, Technical Bulletin, Number 79, July, August, September 1989.

Other Comments: None.

Project:	B. F. Sisk-San Luis Dam	Index No:	PO-027
Owner:	State of California	Feature:	Power Plants
	Department of Water Resources	Stage:	Operation
River/Stream:	San Luis Reservoir	Keyword(1):	Switchyard
Capacity:	440 MW	Keyword(2):	Breakers
Nearest City/State:	Santa Nella, CA	Keyword(3):	Earthquake
Contact Name:	Paul Flanagan		
Contact Title:	Chief of Engineering		
Address:	31770 W. Highway 152		
City/State/Zip:	Santa Nella, CA 95322		
Telephone:	(209) 826-0718		

LESSON LEARNED

Problem: High profile breakers (18-feet + height) could self destruct during an earthquake.

Cause & Effect: Problem caused by earthquake action on high profile breakers.

Investigations: Similar breakers have been damaged or destroyed by earthquakes at other earthquake prone locations.

Analysis/Action: Long-necked insulators were the mark of the SF6-type high voltage circuit breakers in switchyards across the Project constructed in the mid 1960's. Because of the earthquake danger, San Luis Management ordered new breakers. The Japanese-made replacement breakers come in components, for on-site assembly and installation. Civil Maintenance planned to replace Breaker No. 2382 first. Only a few minor problems were encountered so the crew decided to replace two breakers at the same time. The double replacement allowed for shorter outages. Both replacement breakers were assembled in the plant (using the overhead crane) and then were taken to the switchyard. The great advantage in this arrangement was that the crew had the breakers assembled and ready before the outage began. Also, the in-plant dual assembly was a better working and testing environment. On site there was less setup time with two breakers at once. Pouring two concrete pads cut down greatly on the delay of cure time. The first breaker replacement took an outage of six weeks. The next seven breakers averaged less than three weeks each.

Lessons Learned: Do as much work as possible in the plant.

Publications: State of California, Department of Water Resources, Technical Bulletin Number 88, January, February, March 1992.

Other Comments: None.

Project:	Dos Amigos Pumping Plant	Index No:	PO-028
Owner:	State of California	Feature:	Power Plants
	Department of Water Resources	Stage:	Operation
River/Stream:	California State Water Project	Keyword(1):	Wedges
Capacity:	six 40,000 HP Pumps	Keyword(2):	Rotor
Nearest City/State:	Santa Nella, CA	Keyword(3):	Motors

Contact Name:	Paul Flanagan
Contact Title:	Chief of Engineering
Address:	31770 W. Highway 152
City/State/Zip:	Santa Nella, CA 95322
Telephone:	(209) 826-0718

LESSON LEARNED

Problem: Excessive runout (vibration) created a hazardous condition. The investigation disclosed fretting where the shaft and hub of the rotor mated. The keyway had distorted and the keys were starting to roll. The rotor of this unit weighs 143 tons and is 21.5 feet in diameter; the major obstacles included separating hub and shaft from the rim; then, separating the hub from the shaft. The shaft, a solid steel column, 22 feet long, is keyed to the hub of the spider. Then the rotor rim is shrink fitted to the spider and keyed.

Cause & Effect: The motors turn the pump in a CCW direction; upon shut down the 18-foot diameter water column from the penstock stops the pump and turns it in the CW direction in overspeed since there are no penstock valves.

Investigations: Checking the clearance between the shaft and the keys with a feeler guage.

Analysis/Action: Each component was separated from the other by applying heat to the steel surface uniformly so expansion could be controlled to avoid any possibility of deforming the rim, hub or shaft. First, the rim was heated and the hub and shaft (still together) pulled out of it. The hub and shaft were then mounted on another pedestal; heat was applied to the hub and when it was ready, the hub was pulled free of the shaft. At this point, the damage to the keyway and the metal near it was evident. The shaft and hub were rebuilt. The plant personnel aligned the bearing brackets and beefed-up the rotor rim support structure.

Lessons Learned: To check for clearance at keys at regular intervals.

Publications: State of California, Department of Water Resources, Technical Bulletin, Number 81, January, February, March 1990.

Other Comments: None.

Project:	Wanapum Dam	Index No:	PO-029
Owner:	Public Utility District No. 2 of Grant County	Feature:	Power Plants
		Stage:	Operation
River/Stream:	Columbia River	Keyword(1):	Bolt Tension
Capacity:	1000 MW	Keyword(2):	Turbines
Nearest City/State:	Beverly, WA	Keyword(3):	Packing Boxes
Contact Name:	Dave Moore		
Contact Title:	Hydro Civil Engineer IV		
Address:	P.O. Box 878		
City/State/Zip:	Ephrata, WA 98823		
Telephone:	(509) 754-3541		

LESSON LEARNED

Problem: Turbine packing box bolts were failing while in service. Up to half the 60 bolts per turbine had failed. Problem arose after original 1020 D steel studs with raised threads and reduced shank diameter were replaced with grade 5 bolts which are installed into threaded holes without nuts.

Cause & Effect: Problem was caused by inadequate tensioning of bolts compared to the studs and the improper use of bolts for this application. The result was leaking packing boxes, flooded turbine bearings and consequent unit down time.

Investigations: Broken bolt samples were metallurgically analyzed. In place bolts were examined ultrasonically. Original equipment manufacturer was consulted.

Analysis/Action: The investigation revealed the bolts were failing in fatigue. The unit vibration induced enough tension to overcome the initial low bolt pretension resulting in high cycle stress fatigue. A contributing factor was the inability of the bolts to accommodate minor movements and irregularities of the packing box.

Lessons Learned: Bolts should not have been used to replace the original studs. Proper tightening of the bolts was difficult to achieve and bolts, having full diameter shanks and rigid heads did not provide the flexibility of the studs with nuts. All bolts have been replaced with A193 B6 studs with raised threads.

Publications: None.

Other Comments: None.

4.4 OPEN CHANNELS

Project:	Toledo Bend Project Joint Operation	**Index No:**	OD-001
Owner:	Sabine River Authority of Texas and	**Feature:**	Open Channel
	Sabine River Authority of Louisiana	**Stage:**	Design
River/Stream:	Sabine River	**Keyword(1):**	Movement
Capacity:	81 MW	**Keyword(2):**	Surveillance
Nearest City/State:	Burkeville, Texas, Anacoco, Louisiana	**Keyword(3):**	
Contact Name:	Bartom Rumsey		
Contact Title:	Project Engineer		
Address:	Rt. 1, Box 780		
City/State/Zip:	Many, Louisiana 71449-9730		
Telephone:	(318) 256-4112		

LESSON LEARNED

Problem: The intake approach walls are out of alignment with the intake structure pier.

Cause & Effect: The walls were designed and constructed without steel reinforcing across the construction joint between the walls and piers, which allowed the walls to be misaligned by as much as 3 inches at the top of the wall. This unsightly alignment is a concern to the FERC Inspector.

Investigation: Continual surveillance monitoring is ongoing,. To date no additional movement has been recorded.

Analysis/Action: It is assumed that the deflection of the wall is a result of the water pressure from filling of the reservoir. The settlement probably occurred very early in the life of the project during filling of the reservoir and shortly thereafter. The lack of additional movement is attributed to the wall having reached maximum settlement.

Lessons Learned: Construction joint needs to be designed to prevent misalignment.

Publications: None.

Project:	Hatfield Dam	Index No:	OD-002
Owner:	Northern States Power Company	Feature:	Open Channel
		Stage:	Design
River/Stream:	Black River	Keyword(1):	Overtopping
Capacity:	6.0 MW	Keyword(2):	Armoring
Nearest City/State:	Hatfield, WI	Keyword(3):	Drainage
Contact Name:	Richard Rudolph		
Contact Title:	Hydro Administrator		
Address:	100 N. Barstow Street, P.O. Box 8		
City/State/Zip:	Eau Claire, WI 54702-0008		
Telephone:	(715) 839-2486		

LESSON LEARNED

Problem: Power canal dike failed.

Cause & Effect: Local drainage into canal caused overtopping of dike. The dike breached with no significant downstream effects.

Investigations:

Analysis/Action: Dike repair designed with armored crest and slope to resist overtopping flow and/or emergency spillway.

Lessons Learned: Facilities which cut off natural drainage must have adequate provisions for passing or holding the natural runoff.

Publications: None.

Other Comments: None.

Project:	Kootenay Canal	Index No:	OD-003
Owner:	BC Hydro	Feature:	Open Channel
		Stage:	Design
River/Stream:	Kootenay River	Keyword(1):	Debris
Capacity:	529 MW	Keyword(2):	Lining
Nearest City/State:	Nelson, British Columbia	Keyword(3):	

Contact Name:	John Davis
Contact Title:	Senior Engineer

Address:	6911 Southpoint Drive (E07)
City/State/Zip:	Bunnaby, British Columbia, CANADA
Telephone:	(604) 528-2337

LESSON LEARNED

Problem: During the Power plant startup, leakage at the head canal was discovered and construction debris was passing through the turbine.

Cause & Effect: Initial startup water demand exceeded the head canal gate capacity causing high velocities and high pore water pressures. The high velocities caused construction material to be dragged in the intake and the high pore water pressures caused the canal liner to fail.

Investigation: Over the years various investigations have determined how to reduce leakage and repair damage of the powerhouse supply conduit.

Analysis/Action: Divers have placed concrete to contain the remaining debris and seal known leaks in the canal lining.

Lessons Learned: Take the time and assume the expense before the plant goes operational to confirm design velocities and prepare site for startup by removing construction debris.

Publications: None.

Project:	Oswegatchie Development	Index No:	OO-001
Owner:	Niagara Mohawk Power Corporation	Feature:	Open Channel
		Stage:	Operations
River/Stream:	Oswegatchie	Keyword(1):	Winds
Capacity:	0.8 MW	Keyword(2):	Connectors
Nearest City/State:	South Edwards	Keyword(3):	Corrosion
Contact Name:	Jorge Villali, P.E.		
Contact Title:	Licensing Engineer		
Address:	300 Erie Boulevard West		
City/State/Zip:	Syracuse, NY 13202		
Telephone:	(315) 428-5753		

LESSON LEARNED

Problem: Flume failure during high winds.

Cause & Effect: The combination of 50 to 80 mph winds and corroded metal connectors supporting the flume resulted in the flume structure failure.

Investigations: Inspection after failure was performed.

Analysis/Action: Failure was attributed to overstressed metal support connections which prompted inspection of all similar flumes.

Lessons Learned: During design of new structures, avoid steel tie members in locations of difficult access which would hinder easy physical inspection.

Publications: None.

Project:	Bishop 2 Hydro Project	Index No:	OO-002
Owner:	Southern California Edison	Feature:	Open Channel
		Stage:	Operation
River/Stream:	Bishop Creek	Keyword(1):	Dredging
Capacity:		Keyword(2):	Suction
Nearest City/State:	Bishop, Ca.	Keyword(3):	Sediment
Contact Name:	S. F. McKenery		
Contact Title:	Project Manger		
Address:	P.O. Box 800, Bldg GO-3		
City/State/Zip:	Rosemead, CA 91770		
Telephone:	(818) 302-8572		

LESSON LEARNED

Problem: A large volume of sediment had accumulated in the Intake for the Bishop 2 Hydroelectric Power Plant

Cause & Effect: The sediment had to be removed to prevent damage to power plant.

Investigation: The forebay lake was drained and earth moving equipment was unsuccessful in removing the sediment. This operation was not successful due to the high quantity of organic material in the lake bottom which caused the heavy equipment to bog down.

Analysis/Action: A barge mounted suction dredge was successful in removing the sediment. The saturated sediment was pumped to a series of settling ponds where the sediment was settled out and the water was pumped into an irrigation system. Since the water contained suspended fines for a long period, it was not possible to pump the back into the lake.

Lessons Learned: The removal of saturated sediment is best accomplished by dredging. It was also recommended that in the future a rotating cutter head be used on the suction hose to prevent clogging and improve production.

Publications: None.

Project:	Sherman Island Development	Index No:	OO-003
Owner:	Niagara Mohawk Power Corp.	Feature:	Open Channel
		Stage:	Operation
River/Stream:	Hudson River	Keyword(1):	Freeze-thaw
Capacity:	28.8 MW	Keyword(2):	Soil
Nearest City/State:	Glen Falls, NY	Keyword(3):	Insulation
Contact Name:	Jacob S. Niziol		
Contact Title:	Civil Engineer		
Address:	300 Erie Blvd West		
City/State/Zip:	Syracuse, NY 13202		
Telephone:	(315) 428-5556		

LESSON LEARNED

Problem: A small hole was discovered in the concrete power canal slab. The concrete had deteriorated, potentially jeopardizing the integrity of the structure.

Cause & Effect: The 70 year old structure had experienced freeze-thaw deterioration of the concrete progressing from the underside of the slab. If not addressed, this could have lead to the loss of the canal section.

Investigation: The canal was dewatered and the downstream section of the canal was excavated to allow for an inspection of the inside and outside of the canal. Exposed reinforcing steel and loss of concrete was discovered on the downstream section during the investigation. The damaged canal section was supported by buttresses which showed no sign of deterioration. It was also discovered that the buttresses were incompletely backfilled leaving a void between the concrete and embankment.

Analysis/Action: Since the earthfill was not in contact with the concrete no insulation was provided. This allowed a temperature differential with warmer water on one side and colder air on the outside, promoting the freeze-thaw cycles. The remedial action included removal and replacement of the damaged section and restoration of fill so that the concrete was in full contact with the soil.

Lessons Learned: The insulating properties of soil are important for mitigating potential for freeze-thaw damage in thin section structures.

Publications: None.

Project:	Toledo Bend Project Joint Operation	Index No:	OO-004
Owner:	Sabine River Authority of Texas and	Feature:	Open Channel
	Sabine River Authority of Louisiana	Stage:	Operation
River/Stream:	Sabine River	Keyword(1):	Riprap
Capacity:	81 MW	Keyword(2):	Scour
Nearest City/State:	Burkeville, Texas/	Keyword(3):	
	Anacoco, Louisiana		

Contact Name:	Bartom Rumsey
Contact Title:	Project Engineer
Address:	Rt. 1, Box 780
City/State/Zip:	Many, Louisiana 71449-9730
Telephone:	(318) 256-4112

LESSON LEARNED

Problem: The erosion protection material (rip rap) immediately downstream of the powerhouse in the discharge channel was displaced.

Cause & Effect: The discharge flow had scoured the channel and could result in undermining the concrete structure.

Investigation:

Analysis/Action: An outage was taken to reshape the bottom of the channel and replace the rip rap. The channel is now back in service and erosion is being monitored.

Lessons Learned: Need to analysis the discharge flow patterns with a model.

Publications: None.

Project:	Self Cleaning Weir	Index No:	OO-005
Owner:	Pacific Gas and Electric Gas Co.	Feature:	Open Channel
		Stage:	Operation
River/Stream:	NA	Keyword(1):	Sediment
Capacity:	NA	Keyword(2):	Weirs
Nearest City/State:	NA	Keyword(3):	Instrumentation
Contact Name:	Edward Horciza		
Contact Title:	Senior Hydrographer		
Address:	201 Mission Street, P1001W		
City/State/Zip:	San Francisco, CA 94177		
Telephone:	(415) 973-5318		

LESSON LEARNED

Problem: Sediment deposits in gage pools and obstructions of weirs by snow can have a major effect on the weir rating.

Cause & Effect: For many years it has been difficult to obtain accurate flow records for streams with a high level of sediment.

Investigation: Pacific Gas and Electric Gas Co. designed and constructed two types of weirs with hydraulic properties that promote self cleaning in and behind the weir.

Analysis/Action: These self cleaning weirs were installed at five location and have been monitored for the past ten years.

Lessons Learned: A significant improvement in the accuracy of the recorded flow was achieved and much less maintenance and calibration has been required.

Publications: None.

Project:	Kaweah 3 Project	Index No:	OO-006
Owner:	Southern California Edison Co.	Feature:	Open Channel
		Stage:	Operation
River/Stream:	Kaweah River	Keyword(1):	Canal
Capacity:		Keyword(2):	Leaking
Nearest City/State:	Three Rivers, CA	Keyword(3):	Shotcrete

Contact Name:	S. F. McKenery
Contact Title:	Project Manager
Address:	P.O. Box 800, Bldg. GO-3
City/State/Zip:	Rosemead, CA 91770
Telephone:	(818) 302-8572

LESSON LEARNED

Problem: The concrete canal which provides water to SCE's Kaweah 3 Powerhouse was in need of relining, since it was leaking in several places.

Cause & Effect: Repeated freeze/thaw cycles had caused concrete spalling and cracking resulting in numerous leaks. Maintenance costs were increasing.

Investigations: Small selected patching was performed but did not keep up with the spalling.

Analysis/Action: It was determined that a 2" thick, fiber reinforced shotcrete overlay would be applied to the 18,000 foot long canal.

Lessons Learned: The overlay was applied without problems. The canal will be inspected periodically to monitor cracking. A mix design with 50 pounds of fiber per cubic yard of shotcrete was used. A design with 75 pounds of fiber per cubic yard will be used next time to reduce the small spider cracks which developed with this design.

Publications: None.

Other Comments: Stainless steel may not be the best corrosion resistant material for a given fresh water exposure. A thorough investigation is required.

Project:	Lee Vining Substation	**Index No:**	OO-007
Owner:	Southern California Edison Co.	**Feature:**	Open Channel
		Stage:	Operation
River/Stream:	Lee Vining Creek	**Keyword(1):**	Fisheries
Capacity:	N/A	**Keyword(2):**	Creeks
Nearest City/State:		**Keyword(3):**	Gunite
Contact Name:	G. B. Redd		
Contact Title:			
Address:	P.O. Box 800		
City/State/Zip:	Rosemead, CA 91770		
Telephone:	(818) 302-8950		

LESSON LEARNED

Problem: Lee Vining Creek flows adjacent to Edison's Lee Vining Substation, and is subject to high flows during storms. The streambed had been severely eroded, threatening the adjacent substation. The streambed was treated with gunite a number of years ago to stabilize it against flood erosion.

Cause & Effect: The gunited section of creek was successfully stabilized, but a recent agency mandated project to enhance the fishery in the creek revealed that the section was impassable to fish. The gunite caused a smooth channel with increased velocities and provided no pooling for fish to facilitate upstream passage.

Investigations: Excavating the gunite and returning the streambed to its natural state would have been cost prohibitive and would have subjected the substation to erosion damage. Alternatives were discussed with fisheries biologists and engineers. It was decided to install roughness devices in the gunited section to provide suitable passage for fish.

Analysis/Action: Cast-in-place concrete berms and flashboards were installed which formed pools and resting places for migrating fish. The construction work was performed during low flows which allowed water to be diverted around the work. The procedure met with the approval of the agencies concerned. The channel now has a somewhat reduced maximum capacity, but a simple raising of the channel banks with block masonry should correct the problem.

Lessons Learned: Innovation as well as negotiating with other agencies can be beneficial in minimizing costs and reducing operating problems.

Publications: None.

Other Comments: The maximum capacity of the channel will be evaluated during the next high flow period and action taken to raise the banks as required.

Project:	B. F. Sisk-San Luis Dam	Index No:	OO-008
Owner:	State of California	Feature:	Open Channel
	Department of Water Resources	Stage:	Operation
River/Stream:	Off Stream Storage	Keyword(1):	Leaks
Capacity:	440 MW	Keyword(2):	Divers
Nearest City/State:	Santa Nella, CA	Keyword(3):	Grout Holes
Contact Name:	Paul Flanagan		
Contact Title:	Chief of Engineering		
Address:	31770 W. Highway 152		
City/State/Zip:	Santa Nella, CA 95322		
Telephone:	(209) 826-0718		

LESSON LEARNED

Problem: Water leaking through the concrete canal liner; the leak appeared on the surface of the ground some distance from the canal.

Cause & Effect: Water leaked from the canal and intersected with a rodent tunnel. This was caused by settlement of the panels and rodent holes.

Investigations: Investigation included: 1. Monitoring the flows at the point where the water exited from the ground (maximum flow was 75 gal/min). 2. Putting bentonite into a crack that was taking water and observing the bentonite coming out the rodent hole.

Analysis/Action: The lining of the Aqueduct consists of a series of concrete panels cast in place. A leak can develop at the joints or in cracks. Standard repair techniques consist of pumping bentonite or a cement grout behind the damaged liner, down through injection wells; or else pumping through two-inch diameter holes drilled in the bottom of the lining to intercept the path of the leak. Rarely is the Aqueduct dewatered to do such repair work. To pump bentonite down through holes in the bottom of the liner, the first step was for the Mobile Equipment Shop crew to adapt the hydraulic system of an existing rubber tire tractor (20R29) to power the hydraulics of the underwater drill. The tractor's hydraulic system can pump through 100 feet of hydraulic hose reaching the invert of the canal, and operate the rotary drill. After the holes are drilled, the grout is injected using a grout pump powered by a 150 CFM air compressor. This compressor can be moved to the work site with the wheel tractor. The holes in the concrete liner were filled with concrete.

Lessons Learned: Repairs can be made through the concrete invert without dewatering canal.

Publications: State of California, Department of Water Resources, Technical Bulletin, Number 80.

Other Comments: None.

Project:	California State Water Project	Index No:	OO-009
Owner:	State of California	Feature:	Open Channel
	Department of Water Resources	Stage:	Operation
River/Stream:	California Aqueduct	Keyword(1):	Cofferdam
Capacity:	10,000 cfs	Keyword(2):	Canal
Nearest City/State:	Santa Nella, CA	Keyword(3):	Turnouts
Contact Name:	Paul Flanagan		
Contact Title:	Chief of Engineering		
Address:	31770 W. Highway 152		
City/State/Zip:	Santa Nella, CA 95322		
Telephone:	(209) 826-0718		

LESSON LEARNED

Problem: Cofferdam system of constructing a turnout in the California Aqueduct to avoid the cost of dewatering the canal.

Cause & Effect:

Investigations: Compared the economics of dewatering the canal. It was determined that the canal could not be dewatered without miles of the canal buckling from external water and soil pressure.

Analysis/Action: A cofferdam enclosed the working area so that water delivery could continue in the aqueduct and new construction go forward. The cofferdam fitted the configuration of the Aqueduct lining so that the only leaking was around some of the bolts used to hold it against the liner. Fabrication was done at the Panache Water District shop; the cofferdam was then taken to the turnout site. A crane lifted and positioned it. To assure a good seal, 18 concrete blocks (10' x 6') were placed around the top edge of the cofferdam. The lining was cut out in the configuration of the future turnout and earth removed to the correct level. Forms were constructed and concrete poured to create the turnout opening. When the new construction was completed, the gates, controls, and trashracks were installed. Earth was then packed and graded around the new turnout. The final step consisted of removing the protective cofferdam. Merely handling the cofferdam was a king-sized chore. The arms are 32' from side to side; the face is 30' wide; it is 9' high. The Aqueduct was drawn down a little. This made it necessary for some customers to get their water from temporary pumps that lifted over their normal turnout location.

Lessons Learned: This metal cofferdam could be used for all future turnouts up to the size of the cofferdam.

Publications: State of California, Department of Water Resources, Technical Bulletin, Number 81, January, February, March 1990.

Other Comments: None.

Project:	Thomson Project	Index No:	OO-010
Owner:	Minnesota Power Company	Feature:	Open Channel
		Stage:	Operation
River/Stream:	St. Louis River	Keyword(1):	Rapid Drawdown
Capacity:	72.6 MW	Keyword(2):	Dewatering
Nearest City/State:	Duluth, MN	Keyword(3):	Slope Failure
Contact Name:	Stephen Kopish		
Contact Title:	Manager, Production Administration and Hydro		
Address:	30 W. Superior Street		
City/State/Zip:	Duluth, MN 55802		
Telephone:	(218) 723-3952		

LESSON LEARNED

Problem: Power canal dike failed.

Cause & Effect: Rapid dewatering of the canal caused a dike slope to fail due to abrupt unloading. Upstream gates had been closed while plant continued to operate causing rapid lowering of canal water.

Investigations:

Analysis/Action: Dewatering procedures were changed.

Lessons Learned: Review dewatering and drawdown procedures to insure that drawdown rate will be controlled before beginning work.

Publications: None.

Other Comments: None.

Project:	White River Project	Index No:	OO-011
Owner:	Puget Sound Power & Light Co.	Feature:	Open Channel
		Stage:	Operation
River/Stream:	White River	Keyword(1):	Timber-Lined
Capacity:	70 MW	Keyword(2):	Piping
Nearest City/State:	Auburn, WA	Keyword(3):	

Contact Name:	W. J. Finnegan
Contact Title:	Vice President, Engineering
Address:	P.O. Box 97034
City/State/Zip:	Bellevue, WA 98009
Telephone:	(206) 453-6766

LESSON LEARNED

Problem: A section of the timber lined canal failed, disrupting flows to the powerhouse.

Cause & Effect: Exact cause of failure is not known. Apparently, base or side timber collapsed resulting in a cascading failure. Velocity in this section is about 25 fps.

Investigations: Studies recommended replacing the lost section with a trapezoidal, shotcrete lined canal section.

Analysis/Action: The section was replaced but failed when rewatered because of increased hydrostatic pressure and piping of foundation soils.

Lessons Learned:

Publications: None.

Other Comments: None.

APPENDIX A
Questionnaire Form

QUESTIONNAIRE

OWNER DATA

Owner Name: .... >

Contact Name: ... >

Contact Title: .... >

Address: .... >

City/State: ...... >

Telephone No.: ... >

PROJECT DATA

Project Name: .... >

Operating Capacity (MW): >

River/Stream: .... >

Nearest City/State: ..... >

IDENTIFY PROJECT FEATURE (add description beneath)	Check Box for Appropriate Project Stage		
	DESIGN	CONSTRUCTION	OPERATION
Intake/Diversion			
Water Conduits			
Powerplant			
Open Channels			

QUESTIONNAIRE
(use additional pages as necessary)   Page 2 of 3

DESCRIBE LESSON LEARNED

DESCRIBE PROBLEM:

CAUSE and EFFECT (Describe what caused the problem and its impact):

INVESTIGATION (Describe any data acquisition and studies):

ANALYSIS/ACTION (Describe what was done to study/alleviate problem):

LESSONS LEARNED (Describe in detail, with applicability to other projects):

PUBLICATIONS (List references where same problem was previously presented):

ASCE Task Committee
on
**Lessons Learned from the Design, Construction
and Operation of Hydroelectric Facilities**

QUESTIONNAIRE

Page 3 of 3

OTHER COMMENTS (Anything else on lesson learned?):

ASCE RESEARCH and DEVELOPMENT
(List any ideas for future R&D)

Project:	Balsam Meadow Pumped Storage	Index No:	IO-006
Owner:	Southern California Edison	Feature:	Intake
		Stage:	Operation
River/Stream:	Shaver Lake	Keyword(1):	Fish
Capacity:	207 MW	Keyword(2):	Entrainment
Nearest City/State:	Shaver Lake City	Keyword(3):	Cold Temp

Project:	Barkley Power Plant	Index No:	PD-002
Owner:	U.S. Army Corps of Engineers	Feature:	Power Plants
		Stage:	Design
River/Stream:	Cumberland River	Keyword(1):	Spillway
Capacity:	130 MW	Keyword(2):	Chains
Nearest City/State:	Gilbertsville, KY	Keyword(3):	Lubrication

Project:	Bath County Pumped Storage	Index No:	WD-001
	Station	Feature:	Water Conduit
Owner:	Virginia Electric and Power	Stage:	Design
	Company		
River/Stream:	Back Creek	Keyword(1):	Leakage
Capacity:	2100 MW	Keyword(2):	Grouting
Nearest City/State:	Warm Springs, VA	Keyword(3):	Instrumentation

Project:	B. F. Sisk-San Luis Dam	Index No:	PO-024
Owner:	State of California	Feature:	Power Plants
	Department of Water Resources	Stage:	Operation
River/Stream:	Off Stream Storage	Keyword(1):	Amortissuer
Capacity:	440 MW	Keyword(2):	Windings
Nearest City/State:	Santa Nella, CA	Keyword(3):	Motors

Project:	B. F. Sisk-San Luis Dam	Index No:	PO-026
Owner:	State of California	Feature:	Power Plants
	Department of Water Resources	Stage:	Operation
River/Stream:	Off Stream Storage	Keyword(1):	Welder
Capacity:	440 MW	Keyword(2):	Windings
Nearest City/State:	Santa Nella, CA	Keyword(3):	Motor

Project:	B. F. Sisk-San Luis Dam	Index No:	PO-027
Owner:	State of California	Feature:	Power Plants
	Department of Water Resources	Stage:	Operation
River/Stream:	San Luis Reservoir	Keyword(1):	Switchyard
Capacity:	440 MW	Keyword(2):	Breakers
Nearest City/State:	Santa Nella, CA	Keyword(3):	Earthquake

Project:	B. F. Sisk-San Luis Dam	Index No:	OO-008
Owner:	State of California	Feature:	Open Channel
	Department of Water Resources	Stage:	Operation
River/Stream:	Off Stream Storage	Keyword(1):	Leaks
Capacity:	440 MW	Keyword(2):	Divers
Nearest City/State:	Santa Nella, CA	Keyword(3):	Grout Holes

Project:	Big Creek 2	Index No:	WO-006
Owner:	Southern California Edison Co.	Feature:	Water Conduit
		Stage:	Operation
River/Stream:	San Joaquin River	Keyword(1):	Surge Tank
Capacity:		Keyword(2):	Dewatering
Nearest City/State:	Shaver Lake, CA	Keyword(3):	Drainage

Project:	Birch Creek Project	Index No:	WO-003
Owner:	Birch Power Company & Sorenson	Feature:	Water Conduit
	Engineering	Stage:	Operation
River/Stream:	Birch Creek	Keyword(1):	Radiograph
Capacity:	2.7 MW	Keyword(2):	Weld
Nearest City/State:	Idaho Falls, Idaho	Keyword(3):	Testing

Project:	Bishop 2 Hydro Project	Index No:	OO-002
Owner:	Southern California Edison	Feature:	Open Channel
		Stage:	Operation
River/Stream:	Bishop Creek	Keyword(1):	Dredging
Capacity:		Keyword(2):	Suction
Nearest City/State:	Bishop, Ca.	Keyword(3):	Sediment

Project:	Bishop Plant 2	Index No:	WO-004
Owner:	Southern California Edison Co.	Feature:	Water Conduit
		Stage:	Operation
River/Stream:	Bishop Creek	Keyword(1):	Wood
Capacity:		Keyword(2):	Bell & Spigot
Nearest City/State:	Bishop, CA	Keyword(3):	Leakage

Project:	California State Water Project	Index No:	OO-009
Owner:	State of California	Feature:	Open Channel
	Department of Water Resources	Stage:	Operation
River/Stream:	California Aqueduct	Keyword(1):	Cofferdam
Capacity:	10,000 cfs	Keyword(2):	Canal
Nearest City/State:	Santa Nella, CA	Keyword(3):	Turnouts

Project:	Cherokee Falls Hydroelectric Facility	Index No:	PO-006
Owner:	Broad River Electric Cooperative	Feature:	Power Plants
		Stage:	Operation
River/Stream:	Broad River	Keyword(1):	Flooding
Capacity:	82.447 MW	Keyword(2):	Flashboards
Nearest City/State:	Gaffney, S. C.	Keyword(3):	Debris

Project:	Chief Joseph Dam	Index No:	WO-002
Owner:	U.S. Army Corps of Engineers	Feature:	Water Conduit
		Stage:	Operation
River/Stream:	Columbia	Keyword(1):	Baker Coupling
Capacity:	2,069 MW	Keyword(2):	Movement
Nearest City/State:	Bridgeport, WA	Keyword(3):	Leakage

Project:	Colebrook Hydro Project	Index No:	PO-008
Owner:	MDC	Feature:	Power Plants
		Stage:	Operation
River/Stream:	W. Branch Farmington	Keyword(1):	Computer
Capacity:	3 MW	Keyword(2):	Virus
Nearest City/State:	Colebrook, CT	Keyword(3):	Software

Project:	Colorado River Storage - Crystal Dam	Index No:	PD-004
		Feature:	Power Plants
Owner:	US Bureau of Reclamation	Stage:	Design
River/Stream:	Gunnison River	Keyword(1):	Waterhammer
Capacity:	28 MW	Keyword(2):	Vortex
Nearest City/State:	Montrose, CO	Keyword(3):	Venting

Project:	Cougar Dam	Index No:	PO-009
Owner:	U.S. Army Corps of Engineers	Feature:	Power Plants
		Stage:	Operation
River/Stream:	South Fork of McKenzie River	Keyword(1):	Wicket Gate
		Keyword(2):	Galling
Capacity:	2 Units, 12.5 MW each	Keyword(3):	Spatter
Nearest City/State:	Blue River, OR		

Project:	Cuero	Index No:	PO-004
Owner:	Cureo Hydroelectric Partnership	Feature:	Power Plants
		Stage:	Operation
River/Stream:	Guadalupe River	Keyword(1):	Foundation
Capacity:	1.047 MW	Keyword(2):	Settlement
Nearest City/State:	Cuero, Tx	Keyword(3):	Piping

Project:	Curecanti Morrow Point Dam	Index No:	PO-010
Owner:	US Bureau of Reclamation	Feature:	Power Plants
		Stage:	Operation
River/Stream:	Gunnison River	Keyword(1):	Pump
Capacity:	120 MW (2-60 MW units)	Keyword(2):	Mineral
Nearest City/State:	Montrose, CO	Keyword(3):	Deposits

Project:	Democrate Dam/Kern River 1	Index No:	IO-005
	Intake	Feature:	Intake
Owner:	Southern California Edison	Stage:	Operation
River/Stream:	Kern River	Keyword(1):	Drain
Capacity:	24.8 MW	Keyword(2):	· Inoperable
Nearest City/State:		Keyword(3):	Gate

Project:	Dillon Dam	Index No:	PD-005
Owner:	Denver Water Department	Feature:	Power Plants
		Stage:	Design
River/Stream:	Dillon Reservoir/Blue River	Keyword(1):	Turbine
Capacity:	1.75 MW	Keyword(2):	Speed
Nearest City/State:	Dillon, CO	Keyword(3):	Governor

Project:	Dos Amigos Pumping Plant	Index No:	PO-025
Owner:	State of California	Feature:	Power Plants
	Department of Water Resources	Stage:	Operation
River/Stream:	California Aqueduct	Keyword(1):	Keys
Capacity:	15,450 cfs	Keyword(2):	Motors
Nearest City/State:	Santa Nella, CA	Keyword(3):	Rotors

Project:	Dos Amigos Pumping Plant	Index No:	PO-028
Owner:	State of California	Feature:	Power Plants
	Department of Water Resources	Stage:	Operation
River/Stream:	California State Water Project	Keyword(1):	Wedges
Capacity:	six 40,000 HP Pumps	Keyword(2):	Rotor
Nearest City/State:	Santa Nella, CA	Keyword(3):	Motors

Project:	Dos Amigos Pumping Plant	Index No:	PO-022
Owner:	State of California	Feature:	Power Plants
	Department of Water Resources	Stage:	Operation
River/Stream:	California Aqueduct	Keyword(1):	Wear Rings
Capacity:	15,450 cfs	Keyword(2):	Pumps
Nearest City/State:	Santa Nella, CA	Keyword(3):	

Project:	Dos Amigos Pumping Plant	Index No:	PO-023
Owner:	State of California	Feature:	Power Plants
	Department of Water Resources	Stage:	Operation
River/Stream:	California Aqueduct	Keyword(1):	Rotors
Capacity:	15,540 cfs	Keyword(2):	Motors
Nearest City/State:	Santa Nella, CA	Keyword(3):	Laminations

Project:	Ellis Hydroelectric Plant	Index No:	PO-007
Owner:	Arkansas Electric Coop. Corp.	Feature:	Power Plants
		Stage:	Operation
River/Stream:	Arkansas River	Keyword(1):	Corrosion
Capacity:	33 MW	Keyword(2):	Stainless Steel
Nearest City/State:	Fort Smith, AR	Keyword(3):	Pitting

Project:	Falls River Project	Index No:	WC-001
Owner:	Marysville Hydro Partners	Feature:	Water Conduit
		Stage:	Construction
River/Stream:	Fall River	Keyword(1):	Compaction
Capacity:	9.1 MW	Keyword(2):	Hydrostatic
Nearest City/State:	Ashton, Idaho	Keyword(3):	Pressure

Project:	Garvin Falls Hydro	Index No:	IO-002
Owner:	Public Service Company of New	Feature:	Intake
	Hampshire	Stage:	Operation
River/Stream:		Keyword(1):	Debris
Capacity:	17.222 MW	Keyword(2):	Trashrack
Nearest City/State:	New Hampshire	Keyword(3):	Ice

Project:	Glen Canyon Dam	Index No:	PO-011
Owner:	US Bureau of Reclamation	Feature:	Power Plants
		Stage:	Operation
River/Stream:	Colorado River	Keyword(1):	HVAC
Capacity:	1340 MW	Keyword(2):	Access
Nearest City/State:	Page, AZ	Keyword(3):	Platforms

Project:	Glen Canyon Dam	Index No:	PO-012
Owner:	US Bureau of Reclamation	Feature:	Power Plants
		Stage:	Operation
River/Stream:	Colorado River	Keyword(1):	Generator
Capacity:	1,340 MW	Keyword(2):	Rotor
Nearest City/State:	Page, AZ	Keyword(3):	Inspection

Project:	Glen Canyon Dam	Index No:	PO-013
Owner:	US Bureau of Reclamation	Feature:	Power Plants
		Stage:	Operation
River/Stream:	Colorado River	Keyword(1):	Valve
Capacity:	1340 MW	Keyword(2):	Access
Nearest City/State:	Page, AZ	Keyword(3):	Platforms

Project:	Goodyear Lake Hydro	Index No	IO-001
Owner:	Hydro Development Group, Inc.	Feature:	Intake
		Stage:	Operation
River/Stream:	Goodyear Lake, NY	Keyword(1):	Debris
Capacity:	4.585 MW	Keyword(2):	Trashrack
Nearest City/State:		Keyword(3):	Logboom

Project:	Great Falls Hydro	Index No:	IO-003
Owner:	City of Patterson	Feature:	Intake
		Stage:	Operation
River/Stream:	Passaic River	Keyword(1):	Debris
Capacity:	119.292 MW	Keyword(2):	Trashrack
Nearest City/State:	Patterson, NJ	Keyword(3):	Ice

Project:	Hatfield Dam	Index No:	OD-002
Owner:	Northern States Power Company	Feature:	Open Channel
		Stage:	Design
River/Stream:	Black River	Keyword(1):	Overtopping
Capacity:	6.0 MW	Keyword(2):	Armoring
Nearest City/State:	Hatfield, WI	Keyword(3):	Drainage

Project:	Holtwood Dam	Index No:	PC-002
Owner:	Pennsylvania Power & Light	Feature:	Power Plants
		Stage:	Construction
River/Stream:	Susquehanna River	Keyword(1):	Movement
Capacity:	102 MW	Keyword(2):	Reactive
Nearest City/State:	Lancaster, PA	Keyword(3):	Aggregate

Project:	Holtwood Hydroelectric Station	Index No:	IO-009
Owner:	Pennsylvania Power & Light	Feature:	Intake
		Stage:	Operation
River/Stream:	Susquehanna	Keyword(1):	Stoplogs
Capacity:	102 MW	Keyword(2):	Leakage
Nearest City/State:	Lancaster, PA	Keyword(3):	Neoprene

Project:	Holtwood Dam	Index No:	PO-014
Owner:	Pennsylvania Power & Light	Feature:	Power Plants
		Stage:	Operation
River/Stream:	Susquehanna	Keyword(1):	Random
Capacity:	102 MW	Keyword(2):	Vibration
Nearest City/State:	Lancaster, PA	Keyword(3):	Vortex Shedding

Project:	Holtwood Dam	Index No:	PO-015
Owner:	Pennsylvania Power & Light	Feature:	Power Plants
		Stage:	Operation
River/Stream:	Susquehanna	Keyword(1):	Dewatering
Capacity:	102 MW	Keyword(2):	Generators
Nearest City/State:	Lancaster, PA	Keyword(3):	Wheel Pit

Project:	Hoover Dam	Index No:	PO-016
Owner:	US Bureau of Reclamation	Feature:	Power Plants
		Stage:	Operation
River/Stream:	Colorado River	Keyword(1):	Generators
Capacity:	1,935 MW	Keyword(2):	Controls
Nearest City/State:	Boulder City, NV	Keyword(3):	Air Bubble

Project:	Kansas River Project	Index No:	IO-011
Owner:	The Bowersock Mills & Power Co.	Feature:	Intake
		Stage:	Operation
River/Stream:	Kansas River	Keyword(1):	Trashrack
Capacity:	2.1 MW	Keyword(2):	Frazil
Nearest City/State:	Lawrence, KS	Keyword(3):	Ice

Project:	Kaweah 3 Project	Index No:	OO-006
Owner:	Southern California Edison Co.	Feature:	Open Channel
		Stage:	Operation
River/Stream:	Kaweah River	Keyword(1):	Canal
Capacity:		Keyword(2):	Leaking
Nearest City/State:	Three Rivers, CA	Keyword(3):	Shotcrete

Project:	Kootenay Canal	Index No:	OD-003
Owner:	BC Hydro	Feature:	Open Channel
		Stage:	Design
River/Stream:	Kootenay River	Keyword(1):	Debris
Capacity:	529 MW	Keyword(2):	Lining
Nearest City/State:	Nelson, British Columbia	Keyword(3):	

Project:	Lakeport Hydroelectric Project	Index No:	PO-017
Owner:	NH Water Resources	Feature:	Power Plants
	Council/Lakeport Hydro Corp.	Stage:	Operation
River/Stream:	Winnepesaukee River	Keyword(1):	Power Sales
Capacity:	0.705 MW	Keyword(2):	Automation
Nearest City/State:	Laconia, NH	Keyword(3):	Controls

Project:	Lee Vining Substation	Index No:	OO-007
Owner:	Southern California Edison Co.	Feature:	Open Channel
		Stage:	Operation
River/Stream:	Lee Vining Creek	Keyword(1):	Fisheries
Capacity:	N/A	Keyword(2):	Creeks
Nearest City/State:		Keyword(3):	Gunite

Project:	Lewisville Dam	Index No:	PD-006
Owner:	City of Denton, TX	Feature:	Power Plants
		Stage:	Design
River/Stream:	Elm Fork	Keyword(1):	Carbon
Capacity:	2.2 MW	Keyword(2):	Seals
Nearest City/State:	Lewisville, TX	Keyword(3):	Flexing

Project:	Libby Dam	Index No:	PO-018
Owner:	U.S. Army Corps of Engineers	Feature:	Power Plants
		Stage:	Operation
River/Stream:	Kootenai River	Keyword(1):	Runout
Capacity:	525 MW	Keyword(2):	Vibration
Nearest City/State:	Libby, MT	Keyword(3):	Generator

Project:	Lloyd Shoals Hydro Plant	Index No:	PO-003
Owner:	Georgia Power Company	Feature:	Power Plants
		Stage:	Operation
River/Stream:	Ocmulgee, Jackson Lake	Keyword(1):	Flashboards
Capacity:	4.4 MW	Keyword(2):	Corrosion
Nearest City/State:	Jackson, Georgia	Keyword(3):	Leakage

Project:	Lower Monumental Lock and Dam	Index No:	IO-007
Owner:	U. S. Army Corps of Engineers	Feature:	Intake
		Stage:	Operation
River/Stream:	Snake River	Keyword(1):	Uplift
Capacity:	810 MW	Keyword(2):	Drains
Nearest City/State:	Pasco, WA	Keyword(3):	Cold Temps

Project:	Lundy Lake Dam	Index No:	WO-001
Owner:	Southern California Edison Co.	Feature:	Water Conduit
		Stage:	Operation
River/Stream:	Mill Creek	Keyword(1):	Corrosion
Capacity:	3 MW	Keyword(2):	Leakage
Nearest City/State:		Keyword(3):	Spincoating

Project:	Maxwell Kohler, Sunshine, Oradell &	Index No:	PO-002
	Betasso Power Plants	Feature:	Power Plants
Owner:		Stage:	Operations
River/Stream:	Barker Dam & City Water System	Keyword(1):	Gates
Capacity:	3.9 MW (total)	Keyword(2):	Diversion
Nearest City/State:	Boulder, CO	Keyword(3):	Generators

Project:	Murray Hydroelectric Project	Index No:	PD-001
Owner:	City of Little Rock, AR	Feature:	Power Plants
		Stage:	Design
River/Stream:	Arkansas River	Keyword(1):	Sheetpile
Capacity:	36.8 MW	Keyword(2):	Vibration
Nearest City/State:	North Little Rock, AR	Keyword(3):	Connections

Project:	New Martinsville Project	Index No:	IO-008
Owner:	City of New Martinsville	Feature:	Intake
		Stage:	Operation
River/Stream:	Ohio River	Keyword(1):	Debris
Capacity:	34 MW	Keyword(2):	Disposal
Nearest City/State:	New Martinsville	Keyword(3):	Recycling

Project:	Northfield Mountain Pump Storage	Index No:	PD-003
	Project	Feature:	Power Plants
Owner:	Northeast Utilities	Stage:	Design
River/Stream:	Conn.	Keyword(1):	Spherical
Capacity:	1000 MW	Keyword(2):	Valve
Nearest City/State:	Northfield, MA	Keyword(3):	Seals

Project:	Ocoee III Hydro Plant	Index No:	PO-019
Owner:	Tennessee Valley Authority	Feature:	Power Plants
		Stage:	Operation
River/Stream:	Ocoee River	Keyword(1):	Thrust Bearing
Capacity:	28.8	Keyword(2):	Sole Plates
Nearest City/State:	Copper Hill, TN	Keyword(3):	Foundation Bolts

Project:	Oswegatchie Development	Index No:	OO-001
Owner:	Niagara Mohawk Power Corporation	Feature:	Open Channel
		Stage:	Operations
River/Stream:	Oswegatchie	Keyword(1):	Winds
Capacity:	0.8 MW	Keyword(2):	Connectors
Nearest City/State:	South Edwards	Keyword(3):	Corrosion

Project:	Pit 3 Fishwater Release	Index No:	PC-001
Owner:	Pacific Gas & Electric Co.	Feature:	Power Plants
		Stage:	Construction
River/Stream:	Pit River	Keyword(1):	Fish
Capacity:	NA	Keyword(2):	Welds
Nearest City/State:	Burney , California	Keyword(3):	NDT

Project:	Ray Roberts Dam	Index No:	PO-020
Owner:	City of Denton, TX	Feature:	Power Plants
		Stage:	Operation
River/Stream:	Elm Fork	Keyword(1):	Equalizer
Capacity:	1.2 MW	Keyword(2):	Turbine
Nearest City/State:	Aubrey, TX	Keyword(3):	Orifice

Project:	Roanoke Rapids/Gaston Hydro Station	Index No:	PO-021
		Feature:	Power Plants
Owner:	Virginia Electric & Power Co.	Stage:	Operation
River/Stream:	Roanoke River	Keyword(1):	Cavitation
Capacity:	104 MW/225 MW	Keyword(2):	Overlays
Nearest City/State:	Roanoke Rapids, NC	Keyword(3):	Stainless Steel

Project:	Self Cleaning Weir	Index No:	OO-005
Owner:	Pacific Gas and Electric Gas Co.	Feature:	Open Channel
		Stage:	Operation
River/Stream:	NA	Keyword(1):	Sediment
Capacity:	NA	Keyword(2):	Weirs
Nearest City/State:	NA	Keyword(3):	Instrumentation

Project:	Sherman Island Development	Index No:	OO-003
Owner:	Niagara Mohawk Power Corp.	Feature:	Open Channel
		Stage:	Operation
River/Stream:	Hudson River	Keyword(1):	Freeze-thaw
Capacity:	28.8 MW	Keyword(2):	Soil
Nearest City/State:	Glen Falls, NY	Keyword(3):	Insulation

Project:	St. Anthony Falls - Lower Dam	Index No:	PD-007
Owner:	Northern States Power Company	Feature:	Power Plants
		Stage:	Operation
River/Stream:	Mississippi River	Keyword(1):	Sandstone
Capacity:	8.0 MW	Keyword(2):	Foundation
Nearest City/State:	Minneapolis, MN	Keyword(3):	Piping

Project:	Thomson Project	Index No	OO-010
Owner:	Minnesota Power Company	Feature:	Open Channel
		Stage:	Operation
River/Stream:	St. Louis River	Keyword(1):	Rapid Drawdown
Capacity:	72.6 MW	Keyword(2):	Dewatering
Nearest City/State:	Duluth, MN	Keyword(3):	Slope Failure

Project:	Toledo Bend Project Joint Operation	Index No	OO-004
Owner:	Sabine River Authority of Texas and	Feature:	Open Channel
	Sabine River Authority of Louisiana	Stage:	Operation
River/Stream:	Sabine River	Keyword(1):	Riprap
Capacity:	81 MW	Keyword(2):	Scour
Nearest City/State:	Burkeville, Texas/	Keyword(3):	
	Anacoco, Louisiana		

Project:	Toledo Bend Project Joint Operation	Index No:	PO-005
Owner:	Sabine River Authority of Texas and	Feature:	Power Plants
	Sabine River Authority of Louisiana	Stage:	Operation
River/Stream:	Sabine River	Keyword(1):	Erosion
Capacity:	81 MW	Keyword(2):	Draft Tube
Nearest City/State:	Burkeville, Texas, Anacoco,	Keyword(3):	Lining
	Louisiana		

Project:	Toledo Bend Project Joint Operation	Index No:	OD-001
Owner:	Sabine River Authority of Texas and	Feature:	Open Channel
	Sabine River Authority of Louisiana	Stage:	Design
River/Stream:	Sabine River	Keyword(1):	Movement
Capacity:	81 MW	Keyword(2):	Surveillance
Nearest City/State:	Burkeville, Texas, Anacoco,	Keyword(3):	
	Louisiana		

Project:	Toledo Bend Project Joint Operation	Index No:	IO-010
Owner:	Sabine River Authority of Texas and	Feature:	Intake
	Sabine River Authority of Louisiana	Stage:	Operation
River/Stream:	Sabine River	Keyword(1):	Debris
Capacity:	81 MW	Keyword(2):	Logboom
Nearest City/State:	Burkeville, Texas	Keyword(3):	Crane
	Anacoco, Louisiana		

Project:	Upriver Dam Hydroelectric Project	Index No:	PO-001
Owner:	City of Spokane, Spokane, WA	Feature:	Power Plants
		Stage:	Operations
River/Stream:	Spokane River	Keyword(1):	Load Rejection
Capacity:	17.7 MW	Keyword(2):	EAP
Nearest City/State:	Spokane, WA	Keyword(3):	Redundancy

Project:	Wanapum Dam	Index No	PO-029
Owner:	Public Utility District No. 2 of	Feature:	Power Plants
	Grant County	Stage:	Operation
River/Stream:	Columbia River	Keyword(1):	Bolt Tension
Capacity:	1000 MW	Keyword(2):	Turbines
Nearest City/State:	Beverly, WA	Keyword(3):	Packing Boxes

Project:	White River Project	Index No:	OO-011
Owner:	Puget Sound Power & Light Co.	Feature:	Open Channel
		Stage:	Operation
River/Stream:	White River	Keyword(1):	Timber-Lined
Capacity:	70 MW	Keyword(2):	Piping
Nearest City/State:	Auburn, WA	Keyword(3):	

Project:	White River Project	Index No:	WO-005
Owner:	Northern States Power Company	Feature:	Water Conduit
		Stage:	Operation
River/Stream:	White River	Keyword(1):	Wood Stave
Capacity:	1.0 MW	Keyword(2):	Dewatering
Nearest City/State:	Ashland, WI		Collapse
		Keyword(3):	Hoops

Project:	Wissota Hydro Dam	Index No:	IO-004
Owner:	North States Power Company	Feature:	Intake
		Stage:	Operation
River/Stream:	Chippewa River	Keyword(1):	Gates
Capacity:	36 MW	Keyword(2):	Dewatering
Nearest City/State:	Chippewa Falls, Wisconsin	Keyword(3):	Maintenance

INDEX